ATTENDING MARVELS

Attending Marvels
A *Patagonian Journal*

George Gaylord Simpson

With a new Introduction by
Larry G. Marshall
and
a new Afterword by
George Gaylord Simpson

The University of Chicago Press
Chicago and London

The University of Chicago Press, Chicago 60637
The University of Chicago Press, Ltd., London

89 88 87 86 85 84 83 82   1 2 3 4 5

**Library of Congress Cataloging in Publication
Data**

Simpson, George Gaylord, 1902–
   Attending marvels.

   Originally published: New York: Macmillan,
1934.
   1. Patagonia (Argentina and Chile)—
Description and travel.   2. Natural history—
Patagonia (Argentina and Chile)   3. Scarritt
Patagonian Expedition (1930–1931)
4. Simpson, George Gaylord 1902–.
I. Title.
F2936.F56   1982b   918.2 '70461   82-13438
ISBN 0-226-75935-0

". . . All the attending marvels of a thousand Patagonian sights and sounds helped to sway me to my wish."—*Moby Dick*.

CONTENTS

vii

CONTENTS

INTRODUCTION

Attending Marvels is an adventure story. It deals with the anatomy of a scientific expedition, the personalities who participated in it, and the work it accomplished. It follows the day-to-day life of a determined young scientist on the staff of New York's American Museum of Natural History, who, in the early 1930s, ventured to Patagonia, the southernmost part of Argentina, with the definite purpose of amassing an impressive and important collection of South America's fossil land mammal fauna. It gives us an understanding of the pains and labors that went into securing a large fossil collection for one of our nation's great natural history museums, and we share in the anticipated goals, cheer the accomplishments, and despair in the failures that accompany this scientific adventure.

Attending Marvels is also a classic in popular science writing. It is George Gaylord Simpson's first book, published in 1934 when he was 31, and it shows us an important period of his early professional life when he was building the foundations of a remarkable scientific career. In it Simpson demonstrates his diverse talents as a writer as he creates his first major link between scientist and layman. There is nothing else quite like this book for South American paleontology, and for other parts of the paleontological world there are few books its equal. It is also

unique in documenting what human life was like in
Patagonia five decades ago.

In 1975 I visited Patagonia. Armed with *Attending
Marvels* and a copy of Simpson's field notes, I toured
his old haunts and fossil localities. I have returned to
Patagonia several times since and developed a great
affection for the land and its people. My own familiarity with the people and places Simpson knew give me
even greater appreciation of the lasting importance of
Attending Marvels as a literary masterpiece.

After reading *Attending Marvels,* my feeling on
arrival in Patagonia was not surprise, but familiarity.
"I was here before, I know I was, but when? Let's go
find Justino and he will be able to refresh my memory." This familiarity is a gift of the accuracy and vivid
style with which Simpson describes Patagonia's
peoples, places, and things. One needs only the ability to speak Spanish, and equipped with information
in *Attending Marvels* it is possible to discuss local
traditions and customs with natives as if one had experienced them before. *Attending Marvels* has been
read by many Patagonians, and it is not uncommon to
see a copy on the bookshelf of a prosperous *estancia.*
Simpson is warmly remembered by those who knew
him. Others who never met Simpson, but know of
him, speak kindly of "Jorge" as if they were old
friends.

Justino Hernández, Simpson's devoted field assistant, still boasts of his paleontological exploits, and
some of his stories may even be true. During a visit
with Justino at his *chacra* (small farm) south of
Colonia Sarmiento, I asked him if Simpson spoke
Spanish fluently upon his arrival to Patagonia. Justino
assured me that indeed he did not, but that he, Jus-

tino, had taught Simpson virtually everything he knew. To speed Simpson's lessons, Justino reportedly suggested that he write Spanish names on all items in camp (clothes, camp gear, utensils, parts of car, and so forth). Thus, to Justino's self-claimed credit, George Gaylord Simpson learned Spanish.

There is, of course, another side to this tale. George Simpson has told me:

Justino's stories were skirting the truth, not hitting it exactly in the bull's eye. I read a Spanish grammar text on my way down, and I continued to consult it and a dictionary when I had an otherwise idle moment thereafter. Such moments were few, however. I did learn more by conversing with Justino and all the people we met, so I was fairly fluent by the time I returned to Buenos Aires. There some people were amused by the fact that I talked like a gaucho, and occasionally a somewhat obscene one! Carlos Ameghino and I understood each other perfectly, because I had learned his sort of Spanish. Angel Cabrera, however, spoke the most meticulous Castilian, or "Spain Spanish," and conversation with him smoothed out some of the rough spots in my accent and vocabulary.

The story about writing Spanish names on all items in camp is not true, but it has a possible source. In camp I always wore *alpargatas* [a canvas sandal with a hemp sole], which are not shaped for right or left feet but are all the same shape. In order to shape them by always wearing them on the same feet, I marked one *derecha* (right) and the other *izquierda* (left). This amused Justino so much that we wound up labeling the *alpargatas* in as many different languages as we could, including Lithuanian. (You must know that Justino's mother was and his

wife is Lithuanian; "left" and "right" are all I
learned of that language, however.)

George Simpson also told me that Coley Williams,
his other field assistant in Patagonia and preparator at
the American Museum, never did master the Argentine version of Spanish called *Castellano*. When
Simpson and Williams got back to Buenos Aires, one
of Coley's rather numerous English friends there
asked him, "How does it happen that George now
speaks Spanish and you still don't?" Coley replied,
"Oh, George studied it!"

Some things have changed in Patagonia during the
past fifty years. The highway from Comodoro
Rivadavia through the Valle Hermoso and on to Colonia Sarmiento is now paved, and one drives on the
right side instead of the left. The Comodoro-
Sarmiento railway is now defunct, though the tracks
are kept in repair as a national security measure.
Comodoro Rivadavia is now a well-established oil
town with moderate-sized high rises and most of the
conveniences to be found in Buenos Aires. There are
still only three movie theaters, and the bar of the
Hotel Colón is now the best spot in town for meeting
some of Patagonia's local characters.

Away from the paved road and from Comodoro, life
in Patagonia is much as Simpson knew it. The
estancias generally lack electricity; *maté*, the tealike
native beverage, is taken religiously and regularly;
and gauchos are still a foot-loose breed, many
wandering from *estancia* to *estancia* looking for temporary work and avoiding permanent ties. The need
for cooperation and hospitality of local *estancieros* to
get along is indispensable, and Patagonian etiquette is
still rigidly observed.

Although *Attending Marvels* vividly documents the

field season of the first Scarritt expedition to Patagonia (there were two), it records only the beginning of the long and arduous process that makes any expedition a success. There have been many adventurous collecting trips like this one chronicled by Simpson, but it is what Simpson did with the collection and the knowledge he gleaned from it that make his expedition unique. There are those scientists who would have been satisfied simply to have made the collections, remaining content to recount their experiences over dinner tables or around camp fires. But prestige is one form of recognition, and accomplishment is another. Simpson was the complete scientist, and he knew that his adventures in Patagonia marked only the beginning of the work.

A significant aspect of Simpson's contribution to Patagonian paleontology is the meticulous way in which he documented the stratigraphic level of occurrence of the fossils he collected. At some localities the cliffs exceeded three hundred feet in height, with fossils occurring abundantly at numerous levels. Simpson made detailed line drawings of transects taken up these cliff faces, recording the type of sediment and the thickness of each section. When a fossil was collected it was given a number and this number was recorded in the transect. Back at the American Museum it was then possible to reconstruct each transect on graph paper and see exactly where each fossil came from. Paleontologists could then determine whether a particular kind of fossil occurs only (or generally) in one type of rock or in several types. If one kind of animal is found only in sediments representing channel sands and another kind of animal is found only in sediments deposited in a lake environment, then paleontologists can use this information to make inferences about the habits of these animals—

how they lived, and how they may have died. This information also permits paleontologists to document how the faunas preserved in each cliff face changed over time. Did species A from the bottom of the cliff give rise to species M in the middle, and could species M be ancestral to both species X and Y from the top of the cliff?

With these notes in hand, paleontologists can return to the localities where Simpson collected fossils decades before and locate the very level, and often the same spot, from which a given fossil was taken. This information is of great importance, for it permits present-day paleontologists to define problems based on existing collections and then to concentrate their efforts in the field on those levels and at those localities where they suspect the answers may lie. This method takes much of the guesswork out of fieldwork and allows collectors to invest the time they spend in the field efficiently.

This sort of geological information is now regarded as essential by anyone collecting fossils. Yet is was not until Simpson's Patagonia trip that such meticulous attention was given to fossil collecting in the Argentine, and there has not been a comparable concerted effort on such a broad scale anywhere since that time. To repeat this work now would be financially prohibitive. As a result, Simpson's collections and his detailed field notes will long remain the authoritative study of Patagonian land mammal faunas.

The collections Simpson secured were sent to the American Museum, where the long task of preparation began. Many of the larger specimens had been bandaged with flour paste for shipping, a technique Simpson learned while serving as a field assistant to W. D. Matthew in the Texas Panhandle in 1924. These specimens had to be unwrapped and rock carefully removed from around the often delicate teeth and bones. Bones were reinforced, broken bones were glued back together, and each fossil was treated so that it would not be damaged as it was handled during the course of its eventual study.

Once prepared, each specimen had to be catalogued. This involves giving the specimen a catalogue number and placing information about its locality of collection, who collected it, and when, on the collection card. The specimen itself is numbered and identified as accurately as possible. Once catalogued, specimens are placed in storage among other fossils and remain there until scientists want to study them.

Some fairly complete specimens were destined for public display in one of the exhibition halls of the museum. This honor is afforded a specimen that is unique or new to science. In the case of a fossil mammal, the result is typically a skeleton with a skull, mounted in a lifelike pose and accompanied by an artist's reconstruction of what the complete animal may have looked like.

The processes of preparation, cataloguing, and restoration are tedious and require the attention of one or more preparators and a collection manager. If the collection of fossils is large, then the time required may occupy several man years of effort or more. The cost of this work can be staggering. When all factors are taken into consideration, a mounted skeleton of a fossil mammal may have cost an institution thousands or ten thousands of dollars. The market value of the specimen may not be this high, but it would cost a great deal in labor and time to replace it. Space must be made available for the safe storage of collections, and over the years they require attention to ensure that they do not end up in disrepair.

While the cataloguing of his findings began at the American Museum, Simpson remained in Argentina to study the fossil collections of museums in Buenos Aires and La Plata. This work was essential since it permitted Simpson to obtain first-hand knowledge of the animals that had previously been collected, described, and named in many of the same areas he had visited in Patagonia. This work took several months and involved looking through old catalogue records and making careful comparisons of the fossils to ensure that each named specimen was indeed unique. Frequently he found that one kind of animal had been given several scientific names. Simpson found it necessary to determine which name had been given first and therefore had priority of date of publication over the others.

This sort of confusion is not uncommon in paleontology. The fossils collected from a given area are often incomplete, and it is rare to encounter a complete or nearly complete specimen. Thus, a paleontologist studying such a collection may assign a scientific name to a partial lower jaw, another name to a skull part with several teeth, and possibly another name to a foot bone. The eventual discovery of a relatively complete specimen may reveal that the three fragmentary species named actually belong to one species. Then it is necessary for the paleontologist to sort out these "legal" matters. This familiarity with existing collections in Argentina was a necessary prelude to Simpson's later work, for when he returned to the American Museum he was able to identify the specimens he had collected and avoid applying new names to animals named earlier by Argentine workers.

Simpson's work on the American Museum collections began immediately upon his return to New York.

Animals new to science were described, illustrated by technical artists, and given scientific names. These results were published in a series of papers in scientific journals, especially in the *American Museum Novitates* series. Simpson devoted much of his research time during the next thirty-five yars to Patagonian fossil vertebrates, and this effort resulted in scores of papers and in two lengthy monographs based on the collections made from the older beds that he worked. These monographs, published in the *Bulletin of the American Museum of Natural History,* were titled "The Beginning of the Age of Mammals." Part 1 was published in 1948, and part 2 was published in 1967. Information on the collections from the younger beds worked by Simpson have been published in part by Simpson and his colleagues, although monographic treatments have yet to be written. The work on the Patagonian collections thus continues, demonstrating that the description of any large collection is a time-consuming process, one that can occupy a significant part of the career of even such a disciplined worker as Simpson.

In the late 1930s Simpson began to publish his views on aspects of evolutionary theory. He addressed such questions as "Why are there so many kinds of animals?" "Why are certain kinds of animals found where they are?" "How did they get there?" and "How long does it take for one kind of animal (species) to evolve into another?"

It is not surprising that many of the examples used by Simpson to illustrate various points or concepts of evolutionary theory were drawn from his first-hand knowledge of the land mammal fauna of South America. In fact, it was Simpson's demonstrated authoritative knowledge of these and other fossil faunas that gave his theoretical views broad acceptance and high credibility. The Patagonian fossils contributed greatly to the data that formed the foundations of Simpson's theoretical studies as he developed them in such books as *Tempo and Mode in Evolution, The Major Features of Evolution,* and *The Meaning of Evolution.*

Patagonia and its fossil treasures have certainly left their mark on Simpson's career and on twentieth-century science. Indeed, anyone who has experienced Patagonia is haunted by recurrent reflections of it and has mixed feelings toward it. No one who has ever been to Patagonia can forget it. Charles Darwin *(Voyage of the Beagle)* put it this way:

> In calling up images of the past, I find that the plains of Patagonia frequently cross before my eyes; yet these plains are pronounced by all wretched and useless. They can be described only by negative characters; without habitations, without water, without trees, without mountains, they support merely a few dwarfed plants. Why then, and the case is not peculiar to myself, have these arid wastes taken so firm a hold of my memory?

Larry G. Marshall

FOREWORD

"What are you writing, señor doctor?" asked Baliña. This was in his better days, before he had brooded so over Coley's dislike of garlic that he decided to murder us.

"I am writing the happenings of yesterday."

"And why do you write the happenings of each day, señor doctor?"

"That amuses me, and some day I might make it into a book."

"Do you think you could write a book about Patagonia, señor doctor?"

"I don't know. I never tried."

"Pucha! Have you been in Magallanes, or among the glaciers? Have you passed a winter on the pampas? Do you know the imports and exports of last year? Do you know the exact population of Las Heras? Have you spent twenty years here?"

"No. I would just try to write of what I have seen and done and heard myself."

"Pucha! That is not for a real book about Patagonia, señor doctor. I should write a book."

But Baliña went mad and was last seen muttering in his mustachios and heading for the Peeled Lagoon, where the buxom Indian mistress of his heart perhaps awaited him. His book will forever be postponed until mañana, and I have tried to do the best I can without him.

I still do not know the figures for imports and exports, nor have I looked up the exact population of Las Heras. I am sure that the Argentine Government would gladly supply any numerical data that you may desire. I do not care to. This is a narrative and not a handbook. It is a true account,

without statistics, and yet without yielding to the temptation to paint this anomalous and remote region in colors
even more lurid than the original.

When I say "Patagonia" in the following pages, I usually
mean the part of Patagonia that I saw: that central portion
which remained uncivilized for the longest time, left between the two advancing waves of colonization, one coming
north from the Falklands and the Strait, the other south
from the Argentine provinces. In a newspaper account of
our expedition, a helpful editor inserted a map which showed
Patagonia in western Brazil, around the headwaters of the
Amazon. Accustomed as they are to their cold, dry climate
and treeless, wind-swept plateaus, it would be too rude a
shock to transport the Patagonians to the Amazon Basin,
so that the editor must give way and leave Patagonia where
it is. That is in the extreme southern tip of South America,
where it tapers off and points to the South Pole, roughly
between the forty-fifth and fiftieth degrees of south latitude.
This corresponds approximately to North America between
the parallels of New York and central Labrador, and the
climate is rather more severe than that of equivalent latitudes in North America. Comodoro Rivadavia, our Patagonian base, nearly corresponds in latitude to Montreal.

Patagonia is no longer an official name for a country or
any other political division of the earth. The ancient Patagonia has been divided, a narrow strip along the Pacific
belonging to Chile and a wider strip along the Atlantic being
included in the Argentine Republic. The name is still in
common but rather vague use. It is often confined to the
Argentine territories that fall within the limits of historic
Patagonia, and in that sense I shall use it.

This is an account of a scientific expedition, but it is more
concerned with people and events and places than with
science. A definite aim gives meaning and incentive to travel,
but it does not keep voyages to the far corners of the earth
from having interest and excitement not dependent on technical accomplishments. The detailed scientific results of the

Scarritt Patagonian Expedition of The American Museum of Natural History are being published elsewhere, in tomes of interest only to the specialist. This narrative is a less abstruse account of its adventures in Buenos Aires and in the field in Patagonia.

Baliña would not approve. Even though he could not read it anyway, I shall be careful not to send him a copy.

New York.

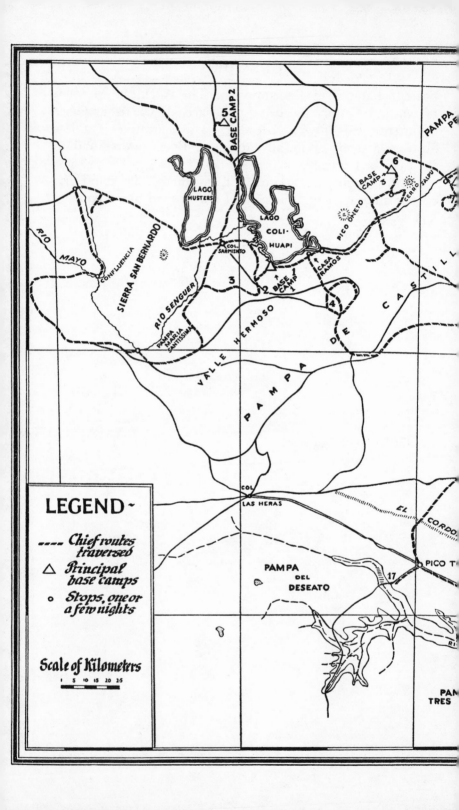

SCARRITT PATAGONIAN EXPEDITION

Chief localities studied in detail:

1 Ameghino's "Barranca al sur del Lago Colhué Huapi"
2 Kilometer 170 of railroad to Colonia Sarmiento
3 Cerro Blanco
4 Near Cañadón Pedro
5 Pajarito, near Cerro del Humo
6 Cañadón Vaca
7 Cañadón Hondo
8 Cabeza Blanca
9 1½ leagues northeast of Cabeza Blanca
10 Lomas Blancas
11 Cerro Redondo
12 Las Tortugas
13 South of Las Violetas
14 Region of Pico Salamanca
15 Cañadón Lobo
16 Punta Nava
17 Pico Truncado

895 miles to Buenos Ayres from Comᵗᵉ Rivadavia to Cape Horn 696 miles

OMALASPINA

BUSTAMANTE

PUERTO VISSER

GOLFO

DE

JORGE

DORO AVIA

FITZROY

FROM PUERTO DESEADO

TO Pᵗᵒ DESEADO

CHAPTER I

REVOLUTION!

THE lecture had not been very exciting. Intimate details about the molar teeth of the larger extinct rodents probably have their place in life, but they are a poor prelude to events more immediate and more stirring. So it was that when Dora, Coley, and I emerged from the subway in the Plaza de Mayo we were ill prepared to have to force our way through a mob of excited young men closely followed by a troop of angry police. We quickly sought a side street, for the Plaza was in the possession of the Escuadrón de Seguridad, the infamous mounted police of the national administration, chasing everyone away with drawn sabers. Their horses were steaming as if they had been working hard, and the police had lost all the geniality of the previous days. The people, or rather the men, for Dora was the only woman in sight, were anxious only to escape.

This was Buenos Aires on Thursday the fourth of September, 1930, and we had stumbled onto the first overt act of the revolution. University students, so often involved in the sanguinary course of South American politics, had started the trouble by parading down the Avenida de Mayo shouting against the government. To keep them out of the Plaza de Mayo, arena of the most violent conflicts in Argentine national affairs, the police had formed a cordon across the Avenida where it opens into the Plaza. The students broke through the cordon, still shouting threats and insults against the government officials who anxiously watched them from the Government House on the opposite side of the Plaza.

A detachment of the Escuadrón was sent to disperse the students. Someone fired a shot. The police, already nervous

1

from the growing tension of the past weeks, lost their heads and charged on the mob. Many were wounded and one of the students, Aguilar by name, was killed. The first casualty of the revolution, he at once became a hero. In the short bitter struggle the police won a brief and misleading triumph. But instead of quieting the unrest, they had only given a martyr to the revolutionary cause and that boy's death in the spring sunshine of the green plaza made civil war inevitable.

We had been too new to the country to foresee clearly what was coming, but this trouble had been brewing for months and some open break had been almost inevitable even before we had landed, eleven days earlier. During those days we had noticed a nightly increase in the numbers of mounted police, standing in groups on strategic corners, or riding the streets, cheerful but armed to the teeth. Aware of the popular feeling against him, and perhaps not distinguished for personal courage, President Irigoyen had shockingly ignored tradition and had refused to open the stock show at Palermo, greatest event of the Argentine year. In his place, he had sent a member of his cabinet and the minister had received such an unfriendly reception from the infuriated populace that he broke and ran, the sheets of his unread speech clasped tightly in his hand.

Next day a government organ published the speech verbatim and recorded it as "read with great effect by the honorable minister at yesterday's inaugural." An opposition newspaper more truthfully reported that "the honorable minister found it inconvenient to stay to watch the march past of the animals. On the contrary, the animals watched the march past of the minister."

The coming storm could well have been foreseen by experienced observers, but we could not believe that such pleasantries would end in drenching these charming and peaceful streets in blood. Still less did we suppose that these incidents directly concerned us in any way.

Our mission was peaceful enough. We had come to explore part of Patagonia and to look for the remains of prehistoric animals there on behalf of The American Museum of Natural History in New York. Our eleven days in Buenos Aires had been passed calmly, familiarizing ourselves with the city, becoming acquainted with Argentine scientists and museums, and arranging for the necessary permits and for transportation to our still distant goal. The harassed government had little enough attention to spare us, and our permits were delayed, but we now had hopes of receiving them soon and leaving the capital.

We had sailed from New York on August eighth on the *Western World*, and the eighteen long, lazy, tropical days of the voyage had passed almost like a dream. After a timeless stretch of blue sky and bluer sea, Pernambuco had floated toward us one morning, a paradise of green trees and gardens studded with buildings of blue, red, and white. There we had our first contact with the tropics and there received that indelible but indescribable impression of which all travelers so inadequately speak. We drove through the crowded streets with their donkeys bearing paniers, their swarthy little policemen, and their two-wheeled carts with old Ford wheels, on past the canals of this "Venice of South America," over their arched bridges, and on into the outskirts where naked brown children played in the mud by thatched huts beneath tall palms.

We stopped, too, at Rio de Janeiro. After dour New York the joyous color of Rio comes with an impact almost physical. That often described scene defies any description and leaves a confused but delicious memory of vivid color and irrepressible life. Brick-red earth, emerald foliage, scarlet flowers, orchids, a cool waterfall in a mass of tangled green, rock grottoes where dinosaurs should roam, the endless, deep blue sea, the unbelievable panorama from the top of Sugarloaf.

In the midst of this, two of our shipmates to whom Fort Worth, Texas, was too constantly present to say that they were from there, spent the entire day in a search for Coca Cola.

Then Santos, squalor and degradation swimming in a delicious aroma of coffee. Thence we drove through the banana

plantations of the lowlands, up the perilously winding road
to the top of the high plateau, and across it to busy, modern
São Paulo. Here on the highlands we first realized that it
was midwinter and that we were getting far south of the
equator.

The lights of Santos were like warmer, friendlier stars as
we slipped silently down the river.

In Montevideo, where the Brazilian motto "Order and
Progress" is more of a fetish than in Brazil itself, we broke a
taxi driver's heart by refusing to get out and look closely at
a nice, new, concrete stadium. He took his revenge by taking
us miles out of our way along a beach, dreary and cheerless
at that season, of which the Uruguayans are unaccountably
proud. Nor did we sufficiently admire "the tallest reënforced
concrete building in the world." Nor yet were we deeply impressed by the fact that Uruguay had the only dollar in the
world worth more than the American dollar—a boast completely shattered soon after. The Uruguayans are more fond
of superlatives than are the most blatant of our fellow-
countrymen.

It was a grand day of fiesta in Montevideo and the town
was blazoned with the flags of the nations, that of the United
States sharing the last and least pole with the red banner of
the Soviets.

And so to Buenos Aires, capital of our new world. Buenos
Aires in September was just emerging from its tempered winter. The parks and plazas were green, but the club-branched
plane trees were leafless and the palms looked valiant but
dejected. Before the revolution started, we had become
familiar with the setting in which it occurred. We had seen
the somewhat pimply, semi-Renaissance, pink stucco Government House, dear to the hearts of the people as the "Pink
House"—Casa Rosada—and had walked from it across the
Plaza de Mayo and down the ten blocks or so of the Avenida
de Mayo to the Plaza Congreso where the new Congress
Building balances and faces toward the Casa Rosada. Like
most of the newer official buildings, the Congress is neo-

classic and of grey stucco. During the revolution shells pierced the hollow columns of these new buildings and exposed their sham of solidity.

We had even learned that Florida is pronounced "Floree'-da" and had walked down this famous street of fashionable shops and cafés during the promenade hour when all smart Buenos Aires strolls here to see and to be admired. In less troubled times the traffic is suspended every afternoon in Florida in order that the fashion show may not be obstructed, but this pleasant custom had been stopped. Florida is a gathering place for mobs as well as for snobs, and traffic was kept circulating through it to try to prevent the formation of inflammable groups. Like all the streets in this old central part of town except the Avenida de Mayo and the two still incomplete diagonals, Florida is a very narrow street, and it runs into the Avenida at right angles near the Plaza.

All themes of Argentine history seem to involve the Plaza de Mayo. Around it arose the first settlement of the now great city. There, in the old Cabildo, part of which is still standing, the nation declared its independence from Spain. The revolution of 1930 started there with the death of Aguilar, and it was destined also to end there.

All through the night after that fatal and stupid incident of September fourth there were riots and skirmishes in the city. The well muzzled papers gave no casualties, but the next morning there were bullet holes in many buildings around the Plaza, and we almost slipped on a pool of fresh blood in Florida.

The fifth was a day of armed truce. In the morning I went to a government bureau, still trying to arrange my affairs with officials who, had I only known it, were to be in full flight in less than thirty-six hours. They probably knew, or strongly suspected, this fact and perhaps were at home packing, for I could see no one. On my way back, I passed the Casa Rosada where a man, bleeding profusely, was being placed in an ambulance and another was lying motionless on the sidewalk. Like so many other incidents of those censor-ridden days, the newspapers did not give even a casual note of this happening.

The guards at the Casa Rosada had been redoubled and every visitor was searched and interrogated before being permitted to enter. In the whole city almost no one seemed to be working. The Plaza de Mayo was crowded with men, a few talking, but most of them standing silently, tensely gazing at the Government House as if trying to divine the intentions of the officials within those walls.

In the afternoon Coley and I went to the National Museum, as we usually did when there seemed no prospect of accomplishing anything in the government bureaus. The building where we were working is one of the oldest in the city, with mud and stucco walls many feet thick, situated on one of the new diagonals only a block from the Plaza. It has been successively a convent, a prison, a barracks, and finally a repository for the bones of the extinct animals that once roamed the pampas where Buenos Aires now stands. Its quiet, cool, and musty old chambers seem a complete refuge from the passions of the world and far removed from the life of today. Shortly after five o'clock, this cloistered peace was abruptly broken by the explosion of several bombs nearby, followed by a great tumult. Our Argentine colleague, the late Dr. Kraglievich, had lived in this atmosphere too long for even this sudden uproar to affect his calmness, but Coley and I ran outside to see what had happened. The streets were full of shouting people, running this way and that as each new rumor spread, of police dashing about and pretending to know what was going on, and of harassed shopkeepers defending their windows and putting up their steel shutters.

Memories of the revolution are inextricably associated with sounds. The punctuating noises, and the dreadful silences, of those days seem to come more vividly to mind than do the actual scenes. And even the thud of field guns, the drumming of machine guns, and the roar of crowds seem somehow less characteristic than the incidental sounds less noticed at the time. Among these, that clangor of steel shutters stands out keenly, and with it the coughing whistles of the exhaust horns on the ambulances.

After gathering a dozen conflicting reports, we learned that the bombs had been less martial than they sounded. They had been set off harmlessly as signals of important news: President Irigoyen had handed over the government to the Vice-President, Enrique Martínez, and the first act of the latter had been to suspend the constitution and to declare the city to be under martial law. This was the last desperate effort on the part of the government to avert armed rebellion.

That night our block was closed to traffic and a reserve troop of mounted police, the dread Escuadrón de Seguridad, "Irigoyen's Cossacks," was stationed there. Yet I find in my journal "And so, with Irigoyen's resignation, the threatened revolution has perhaps ended." I am no prophet.

With ever changing political opinion, September sixth may not remain in Argentine schoolbooks as a major date, but within the span of my stay in the Argentine it seemed certain to become so. It was the "glorious Sixth." Tangos were named for it. Postage stamps were issued to commemorate it. Innumerable books attempted to explain it. A monument was started to celebrate it. Being present at the original Seis de Setiembre was rather like being present at the original Fourth of July. We saw a national holiday in the making.

The morning of the sixth the streets seemed quiet. The newspapers gave no hint of a revolution, actual or impending, and even their front pages were chiefly devoted to beauty hints, weather reports, and useful suggestions as to the best way to mend broken china. Confirmed in the belief that all was over, I went out with Coley to transact some business. As we walked up the Avenida de Mayo, everyone around us suddenly started running. We stopped and looked around until Coley said "At least let's get behind a tree." So we solemnly stood behind a sapling about two inches in diameter, but nothing happened. The alarm was only evidence of the taut nerves and mob psychology that reigned on that day. We found that we could get no business done because everyone either was outside rumoring and rioting or else had gone home to be out of the way of what might, and did, happen.

As we turned homeward a squadron of fighting planes appeared in the sky, giving the earth-bound in the city tightened throats as they realized that before this menace they were no better than rats in a trap. The news rapidly flew from mouth to mouth that an army was marching on the city from a large encampment in the suburbs. At first it was naturally assumed that these were loyal troops coming to strengthen the government and to enforce martial law, but as we ate an early lunch more definite and more ominous word was received. The troops were revolutionary, coming as invaders to capture the city and to wrest the government from the partisans of Irigoyen if possible. Immediately after lunch I went out again.

In the streets there was an awesome silence, as if the city were holding its breath and awaiting some dreaded event. Not a woman or child was in sight, only here and there small clusters of men, grimly speechless or talking in low, nervous tones. I tried to enter the Avenida de Mayo along Florida, but encountered members of the Escuadrón who waved me back without a word. On the next street I was turned back still more decisively, but finally by making a wide detour I managed to slip into the Avenida. Here there were more men. I walked a few blocks toward Congreso, and then stopped on a corner to talk with some other civilians. We were quiet, and none was displaying any arms.

A block away a troop of Irigoyen's cossacks started toward us. After the incident of the fourth there had been some witticisms to the effect that the Escuadrón always blew a bugle three times before they killed anyone, so when one of these raised a bugle to his lips, I prepared to go elsewhere. But before he could blow, the officer, either a poor sport or ignorant of the rules, gave a sharp command and the troop opened fire on us. A man beside me fell screaming. I did not delay.

The shouts and firing redoubled behind me. Some of my

fellow strollers had been less innocent than they appeared and were returning the fire of the police. Being unarmed and disinterested, I sought shelter, and finally came to a door which had been left ajar. I crashed into it, wedged my way in, and closed it behind me. I was with several other fugitives in a tiny clock shop, where numerous clocks of all sizes ticked peacefully, each telling a different time. The proprietor was an old, old lady who was quite unaware of any reason for this invasion and who babbled and cursed us feebly until one of the party took off his hat, made a low bow, and gave a little speech.

"Señora," he said, "I am deeply sorry that we have disturbed your peace. We are gentlemen, all of us, and ordinarily we would never be so rude to so charming a hostess, but stern necessity impelled us. Outside there are a number of friends of the Peludo [Irigoyen] who desire to kill us, and we do not approve of their plans. Your unwilling hospitality has saved our lives. On my own behalf, and on behalf of all these gentlemen, your servant begs that you pardon us and gives you one million thanks."

Somewhat placated, she subsided to mumbling behind toothless gums. The shooting directly outside stopped after a time, and we left, one by one. I had not gone a block before more horsemen appeared at the corner and began using their sabers on the stray civilians. One man, like a frightened rat, ran from door to door, trying them in vain. At last he was overtaken, a saber flashed and he fell. Perhaps the slaughter of non-combatants that day had some purpose or excuse which I, as a foreigner, could not appreciate, but to me it looked like utter brutality of the mounted police, most of whom are Indians, freed of restraint by the revolution and bound to taste blood before the régime that supported them fell.

More fortunate than some, I found an open door and entered the lobby of an apartment building.

"This is all the *Crítica's* doing," said a fellow refugee. The *Crítica* is, or was, a yellow journal patently modeled on some of our own.

"Yes," said another, "It all comes of North American ideas."

"Truth. It is because of the Yanquis that they are killing us."

"It is all a plot of the North American imperialismo!"

I was already accustomed to hearing my country abused and made the scapegoat for all the ills that beset Latin America, but this conversation began to make me feel a little self-conscious. I resolved to keep quiet in order not to betray my accent, which must have been particularly terrible then when my Spanish was still decidedly in the formative stage, and I left as soon as possible.

After another detour, I went back into the Avenida de Mayo. This was the worst tactical error of the day. Almost at once a machine gun opened up very close by. It was answered by another, and then came a perfect fusillade of machine guns, rifles, and revolvers. I had, as I later learned, walked right into the major engagement of the day. Even then it was apparent that I had not yet found the proper atmosphere for quiet study and meditation. I departed, rapidly, running down the Avenida in search of shelter. Luck still held, for there was a steel shutter not tightly closed and leaving about two feet between it and the sidewalk. I squirmed under and the proprietor swore at me and slammed the shutter down and locked it. Someone opened fire immediately outside, where I had been standing. There was a sound of running feet and someone pounded on the door.

"Amor de Dios! Let me in. I am dying."

"No. Go away."

The shutters gave a false sense of security. Next day many were seen to be riddled by bullets, but our shutters either were not hit directly or were of stouter stuff.

This time I was in a café where a number of refugees were passing the time pleasantly drinking beer. I did not hesitate in joining them. Most of them were cheerful but excusably nervous.

"Soy hombre de paz!" insisted one. "I am a peaceful man!

Why do they shoot at me? My nose, only my nose I stuck out and paff! someone must shoot at it. At me, a peaceful man! Why is everyone so mean to me?"

Another spent his time at the telephone.

"Central, central! The slut, she isn't there. Ah, central, who is dead? What was the shooting in the Congreso? What army is this outside? What?—

"Gentlemen, friends, the general is dead! Shot! Oh, no. I am mistaken, the general is at Luján! There, I have lost the connection again. Central!"

"You can't learn anything there. Let me telephone. I must tell my family I am alive. My family will be in hysterics. We were going for a picnic. Let me have it."

"No use. They won't give you a connection."

"Mozo! Two more beers!"

"There goes a field gun at Congreso! They will shell the building! They will burn us! Señor proprietor, let us out before we are killed here!"

"No, señores! I will not open for anyone. You came here, here you stay!"

"Let us up to the roof. We will go over the housetops."

"No. Here you stay!"

"Three more beers!"

"Madre de Dios! That was a bomb."

"Shut up! Mozo, bring dice! Let's play *cinco cosas*."

"All right then. Mozo! Four more beers!"

Only one man was quiet, sitting in a corner trembling violently, his face the color of putty.

Battle raged outside. Finally the proprietor let us up to the second floor where we could look out over the balcony. Things were getting quieter. A building was burning at the Plaza del Congreso nearby and sporadic shots still came from there, but the field guns and machine guns soon stopped.

Earlier incidents had been futile and senseless. The street fighting had been between civilians and the Escuadrón de Seguridad. Many of the civilians were well armed and gave a good account of themselves, but they had no useful rôle in the revolution. The present engagement was the only one that had any military significance. At the moment when I reached the Avenida, cadets from the Argentine West Point, forming the vanguard of the revolutionary army, had marched into the Plaza de Congreso. They had innocently walked right into an ambush. No sooner were they well into the Plaza than concealed supporters of Irigoyen opened fire from all sides with machine guns. Theoretically the cadet corps should have been wiped out, but thanks to the very poor marksmanship of the enemy and to the remarkable coolness of the cadets, they had amazingly few casualties. They immediately deployed, returned the fire, and then charged the buildings from which they were being fired on. They cleaned them out, burning several.

This was the only organized opposition to the invading army. The government had been relying largely on the fleet, which was in the harbor. The navy remained sufficiently loyal not to join in the revolutionary movement, but it refused to fire on the army or to take any active part in the defense of the Irigoyen régime.

After a time the shooting stopped altogether. All the Irigoyenistas had either been killed or taken to their heels. We all went outside into the Avenida again. The streets were slowly beginning to fill. The brutal Escuadrón had disappeared as if by a miracle, except for one small troop that had gone over to the winning side as soon as they saw how the wind blew. They were soon surrounded by a friendly mob, apparently quicker than I to forget their rancor against these bloody half-castes. Within half an hour the Avenida was jammed with people.

Suddenly a white flag appeared on the Casa Rosada. The government had surrendered! A great shout arose and swept down the Avenida like a flood. There was a mad rush for the surrendered government house and it was soon full of jubilant citizens, who filled every office and balcony and even crowded the roof. The whole population seemed to have appeared by magic and to have gone crazy with joy. The shouts were deafening. Sirens and whistles were blowing, friendly bombs bursting in the air. Students commandeered busses and drove them around on the sidewalks and paths of the Plaza de Mayo, shouting the national anthem at the tops of their voices.

A shrieking car came pushing through the mob bearing two officers, Uriburu and Justo, leaders of the revolution. Men crowded around them, cheering, slapping them on the back, kissing them. They entered the Casa Rosada where the only member of Irigoyen's administration who had the courage to stand at his post sadly handed over the government to them. The revolution was an accomplished fact.

While the yelling mob was surrounding the scene of his triumphs and demanding that he be hanged, Irigoyen, a broken old man, was actually driving through the streets unrecognized. He drove to La Plata and there surrendered to the army. He was eventually placed on a small island in the river and kept prisoner there for over a year, when another change of government finally released him.

Politics are a dull business unless mixed with guns and sabers in the South American manner, but Irigoyen's rise and fall was a fascinating romance even aside from the struggle that ended his political career. Leader of the majority party, the Unión Cívica Radical, he had attained a tremendous personal following and had served a previous term as president. Since a president cannot serve two successive terms under the constitution, Irigoyen had placed Alvear in the chair to succeed him. Intended merely as a figurehead to fill in the time until Irigoyen could be president again, Alvear had developed alarming independence. He split the Unión Cívica Radical into two fractions, the Irigoyenistas or Personalistas and the Antipersonalistas. At the next election there was a bitter contest, but the Personalistas carried the day and Irigoyen was elected for a second term.

No sooner was he back in power—this was in 1928—than the people found that they had saddled themselves with an old man, not entirely sane, and wholly incompetent. He had visions, and guardian angels advised him how to run the country. He lived in moody, almost subterranean, isolation—hence his nickname of "Peludo," the peludo being a native animal, a species of armadillo, of unclean habits and given to burrowing and nocturnal prowling. His closest associates were apparently a bootblack of Italian descent and political leanings and two young ladies whom he brought to the capital from a provincial town and placed in government posts. He brooked no interference in his divinely dictated personal plans, yet he interposed no check on the corrupt activities of his rapacious followers.

So in 1930 the Argentine found itself in desperate condition, driven to the verge of ruin by dishonest rulers and headed by a doddering visionary with four years more to serve. Of South American countries, the Argentine has shown the least tendency to have recourse to arms in its political affairs. It does not have the revolutionary habit that we are inclined to impute to all of Latin America indiscriminately. But here it was faced by a situation that, in the view of many of its best citizens, was soluble only by an appeal to arms. The revolution of September Sixth was the result.

The evening of the sixth was one of public rejoicing and private violence. Men broke into taxidermists' shops, took some stuffed peludos (the animals) and hanged them to lamp posts in a gruesome pun. Busts and statues of Irigoyen suffered similar indignities. One, made of quebracho wood, was hauled through the streets with ropes and then burned in a gutter. The presidential chair was carried out of the Casa Rosada. A mob of students set fire to it and carried it through the streets, shouting "Se acabó!"—it is finished! That phrase was chanted on every side. Irigoyenist partisan newspapers were looted and their buildings burned. Opportunists seized the chance to settle private grudges and there was some shooting, but the general air was one of holiday and jubilation. All the buildings around the Plaza de Mayo were outlined in electric lights, a favorite display for all holidays, very pretty but not art.

The following day, Sunday, was devoted by Señor and

Señora Buenos Aires to admiring the numerous bullet holes in trees and façades. This time everyone thought that the trouble was over, and no one suspected that the most tragic episode of all was still to be enacted.

Monday, the eighth, was proclaimed a national holiday. In the afternoon the provisional (revolutionary) government was sworn in on a balcony of the Casa Rosada. This ceremony was witnessed by a crowd that would be the despair of a fight promoter and the envy of a football college. At least 300,000 people were present, and most estimates were higher. The great Plaza de Mayo was jammed with people, leaving hardly room to breathe. Every balcony, window, and roof that afforded even the most distant view of the ceremony was occupied. The leafless plane trees blossomed with small boys. Everyone was gay and in festive mood. Someone released flocks of pigeons, white and with their wings dyed blue. Thus adorned in the national colors, they circled overhead throughout the proceedings.

South America is a land of volatile emotions.

During dinner, we suddenly heard machine and field guns being fired nearby. This seemed too extreme to be a simple expression of high spirits and we ran out on the balcony. Street car crews had abandoned their cars, and passengers and pedestrians were running for shelter. We went down to see what the trouble was and started walking toward the Plaza de Mayo. A man came running up, shouting "Cuidado con la Plaza!"—"Look out for the Plaza!" We stopped him and he told us that a counter-revolution had started. He was a reporter for an English-language paper and had run the gantlet of fire to a cable office only to find that rigid censorship had been imposed and no news of the revolution or supposed counter-revolution could be sent.

We went on, stopping at every corner to be sure we were not running into a barrage. The Plaza and adjoining streets were swept by machine guns.

Only a few hours earlier, three hundred thousand rejoicing people had packed the Plaza. Now the buildings were still gay with their festoons of lights, but they lit a scene of terror and death. The Plaza itself was almost deserted. The only vehicle in it was a derelict bus with all its windows shot out. A few men with rifles lurked behind trees, occasionally dashing, hunched over, from one shelter to the next. Other armed men stood behind the columns of the cathedral. Most of the firing was coming from the direction of the Casa Rosada. The bullets made an odd noise as they chipped the polished granite bases of the buildings. Some wounded or dead were lying on the pavements. From time to time an ambulance, with a man on the running board waving a white flag, would go shrieking across the Plaza and return loaded.

Civilians near me had broken into an arms shop and taken weapons and ammunition. They were now happily engaged in setting fire to the shop. A dignified old gentleman in formal dress and with a neat grey beard came walking up. As he passed me he calmly pulled a large revolver out of his pocket and went on into the Plaza. Following in his footsteps later, I came to a line of holes in a façade—machine-gun fire—and in front was a pool of blood and a torn, red-stained formal collar.

At the corners of all the streets debouching on the Plaza were groups of civilians armed with everything from machine guns to pocket knives. None of them seemed really to know what was going on. The men near me thought that the Casa Rosada had been captured by counter-revolutionaries, and they were preparing to recapture it. To find out, I hopefully held up a white handkerchief as emblem of my non-combatant status and went down to the Casa. It was still in the hands of the revolutionaries, who were, as they thought, engaged in beating off an attack by the Irigoyenistas. In such circumstances, those most actively engaged know the least about the real course of events. Fighting soon stopped there, and I went back up to San Martín, the street by which I had entered the Plaza, planning to go down it in search of Dora and Coley, who had stayed there.

As I reached the corner, there was a burst from a machine

gun about two blocks down San Martín. A young civilian blew a blast on a bugle, waved a flag aloft, and started running down the street, crying "Forward, comrades!" After running a hundred yards he suddenly discovered that no one was following him. He dropped the flag and sheepishly retreated, even more rapidly than he had charged. Someone came from the direction of the firing and told us that all was well. It was only a civilian who had looted a machine gun from an arms shop. He was trying it out by firing it down the street, simply to see how it worked and in a spirit of good clean fun. As soon as he stopped, I went on.

A few blocks down San Martín two young men obviously in a very excited state stopped me. It seemed best to stop. Grimy hands were on my shoulders, the muzzles of two rifles were pressed hard into the pit of my stomach, and my assailants were eagerly fingering the triggers.

I wish I could say that I looked the lads coldly in the eye and told them in flawless, fluent Spanish to go on about their business. I did nothing of the sort. I fully expected to die immediately and messily, and the thought almost paralyzed me. In most perilous situations, death does not really seem imminent and inevitable. There always seems a chance, or there is always something to do. The feeling that now struck me that there was really nothing to be done and no time to do it in was, I think, the most unpleasant emotion I have encountered. It did not even occur to me to put up my hands.

It was probably only a few seconds, but it seemed like an age before I managed to say in halting Spanish that I was a foreigner, not involved in their political battles and was on my way home. They hesitated, then whirled me around and jammed their rifles into my back. I meekly walked away, my spine prickling horridly. At the next corner I turned and by going around the block got back onto San Martín again. My late aggressors had disappeared, and I looked about for Coley and Dora but could not find them. When I got home I found that they had returned an hour or so before and were safe and sound. The indentations of the rifle barrels, still on my coat, suddenly gave me an unsettled feeling, but a stiff whiskey fixed that up and another cured Coley of the shock of hearing of my adventures.

And all that murderous evening had been a mistake. The combatants were all on the same side. Troops had been posted in the Post Office, in the Customs Building, in the Casa Rosada, and in the Plaza Colón, within sight of each other but not in direct touch. Just after dark, a small band of supporters of the old régime drove rapidly past, firing revolvers into the air. Each body of revolutionary troops thought itself attacked, and they blazed away at each other and at everyone within sight. The actual protagonists drove away into the night and none was discovered until a year later. Thinking that the revolution was in danger and that the buildings were captured by Irigoyenistas, the citizens also arose and joined heartily in the mêlée. As they attacked the revolutionary troops, they were themselves taken for counter-revolutionaries. The whole tragedy of errors became so involved that most of the combatants just settled down to shooting at anything that moved, and it was several hours before they found that there was no enemy.

Next day the town was plastered with posters reinforcing the edict of martial law and proclaiming that anyone caught in any act of violence would be shot forthwith. The provisional government highly praised the spirit of the citizens who had, or thought they had, risen to its defense, but it was prudently ordered that all civilians must be disarmed by nightfall, on pain of capital punishment. Soldiers policed all the streets. So peace was restored and the revolution was really at an end. It was a nice revolution: just enough to be typical, and not enough to be boring.

The change of government caused some delay in our affairs. It was necessary for the new ministry to approve our permits to explore and collect in Patagonia, and much of our previous work in this direction had gone for nothing. Two weeks more were devoted to this, while in our spare time we worked at the Museo Nacional and visited La Plata and its excellent museum. On September twenty-third, Dr. M. Doello Jurado, who had been most gracious and helpful to us throughout, assured me that we could rely on the permits being granted, even though they were not yet actually in hand. Coley was then very ill and could not travel, but I decided to delay no longer. I would go on to Patagonia while he stayed in Buenos Aires in his wife Dora's care until he had recovered. Dora would then return to the United States and Coley would rejoin me in Patagonia.

CHAPTER II

ENTRANCE TO PATAGONIA

THE *Ministro Frers*, government oil tanker bound for central Patagonia, lay in the Dársena Sud of Buenos Aires in the wet, grey dawn of September twenty-fourth. A bored taxi driver, two volunteer guides, a large stack of luggage, and I threaded our way painfully and with several false turnings through the maze of shipping, and finally located the tanker tucked away among many more imposing vessels.

This was a very unimportant sailing, hardly noticeable in the general hubbub of the docks. An itinerant shoe-mender strolled up hopefully with a last, a string of heels, a roll of leather, and a box of tools attached to various parts of his person, but his shout of "Zapatero" went unanswered, and he wandered disconsolately away. The picture-while-you-wait man had more success; a farewell party managed to interrupt embraces long enough to pose self-consciously for a tintype of the occasion. Another misanthropic merchant arrived and fought his way on deck with a pile of magazines, apparently salvaged from gutters, and a basket of dejected fruit. He seemed surprised and even a little hurt when he made a sale. I bought a copy of *Caras y Caretas*, destined never to be read.

A smart young army officer stood on the dock, tearing his hair and calling on his patron saint. His automobile, a roadster painted the most virulent orange imaginable, was being slung on board, and the playful stevedores had succeeded in getting the tackle thoroughly snarled and in denting all the fenders of the car, without any sign of really getting it loaded. When he could no longer bear the sight, he rushed back to a large group of women who embraced him

20

and wept noisily. My own truck had been placed on board some days before, and was on the forward deck, heavily loaded with equipment and covered with a tarpaulin, tightly lashed at the corners.

We were due to sail at 9:00 A.M. By 9:45 the last farewell was said and the female relatives of the young officer stood on the dock wailing. At 10:30 all seemed to be ready. At 10:45 the whistle blew and the gangplank was taken up. At 11:15 they started to cast loose, and at 11:30 the tanker was pulled from its berth and towed carefully through the crowded dársena, only two and a half hours late and so rather sooner than was normally to be expected.

We had scarcely left the harbor and were fighting our way against a head wind down the great River Plate when lunch was announced. My still incomplete knowledge of Argentine diet tricked me, for after four or five courses of only moderate appeal I began to fear an unsatisfied conclusion and filled up on puchero, which is food only for the very hungry or the very native. More or less in the manner of that other abomination, the New England boiled dinner, it is made by throwing several kinds of meat and any vegetables that may be on hand into a kettle of water and boiling them until the last trace of flavor has been extracted and is poured off with the water. This juice makes excellent soup, but the residue, the puchero, tastes like moist excelsior and is hailed by the patriotic criollo as a triumph of cookery. On this insult to Savarin I sated my appetite, and then sat by in impotent jealousy while my table mates devoured courses of stewed kidneys, fried eggs, grilled meat, fruit compote, fresh fruit, and cheese. Thereafter I expected the cook to display his entire repertory at every meal, and learned to choose more wisely.

An oil tanker is a one-purpose vessel unlike anything else that sails the seas. It has an atmosphere, and an odour, all its own. Primarily it consists of a series of large tanks, with an exposed complex of business-like but mysterious valves, manholes, and pipes. Forward (on the *Frers*) is a small

structure for steerage passengers, all of whom were workers
going to the oil fields, and aft are the engines and the crew's
quarters. Nearly amidships is a small structure with the
bridge and quarters for the officers and first-class passengers,
a few staterooms and a central salon which is dining hall, bar,
reading room, and promenade rolled into one. On the forward deck was some solid freight: a stack of well casing, our
expedition truck, and, incongruously bound for wild Patagonia, a modern street-sprinkler.

The cabins were not all occupied, but for company they
placed me with another American. For thirty-six hours I saw
nothing of my fellow but an amorphous bulge in the opposite
bunk. When he had finally recovered from his farewell to
Buenos Aires, he emerged slowly and turned out to be a salesman, hardy breed that represents our nation (not always
unfavorably) in every corner of the globe. He was as little
shy as most of his colleagues, and to maintain a conversation
with him it was necessary only to stay awake and to nod
from time to time. Indeed, it was not always necessary, or
possible, to stay awake. There were not enough true things
in the world to sustain that flood of speech, and he was one
of the most truly accomplished liars that I have ever met.
To hear him was a complete education in that gentle art.
With true temperament and love of lying for its own sake, his
complicated prevarications were so embellished and so grandi-
ose that they deceived no one and entertained all.

The whole company met for meals. Our small group included two ladies, one placid, plump, and silent, the other
energetic, thin, and voluble. The latter used to tell long, pointless tales in Spanish, fluent but with an accent strange even to
my unpracticed ear. One day she broke into English and
announced her name as Kelly, a common Argentine name but
in this case not so far removed from its native sod as was
Kelly the dock laborer whom I met in Buenos Aires. He had
only three words of English but those he spoke with a thick
brogue inherited from his Irish grandfather. He called himself Kay-zheé.

"Ghosts are very common in Patagonia," said the Señora

Kelly one day. "Often at night I hear horsemen riding past our estancia. One night I ran out and saw one. He was all in grey and said never a word. Then he disappeared, pouf! In the morning there were no hoofprints. It was not a man. It was Death, returning in the storm from a grisly errand on the pampas."

"And why not?" said the company. "That can well be!"

Dear old Señora Kelly. I wish I knew where she is now; she owes me fifty pesos.

Then there were two nondescripts. I never learned their names, or conditions, but that one was a person of importance was clear from the gold toothpick that he carried in his vest pocket and from his refined way of concealing his operations with it behind a napkin. The two nondescripts played chess together, and so were beyond the reach of human society. Yet not quite always so, for one day at table Gold-toothpick and the young officer somehow became involved in an argument over the value of the army.

As an esthetic performance, honors in the dispute undoubtedly rested with the officer. Gold-toothpick gestured chiefly by popping his eyes in and out, while his antagonist used every muscle from head to foot. The most graceful finger motions were combined with shrugs of all grades, from little "well, well" shrugs to great "be damned to you" shrugs which seemed to shake the whole subject off his shoulders and into the soup. His head nodded, his eyes flirted and glared, his eyebrows tipped, and all the while his delicate fingers went on ticking off regiments, reducing things to tiny atoms (forefinger on thumb), scattering benefits, or chasing out malefactors, all attuned to the explosive obbligato of his voice. Finally—

"Señor, you have insulted the army!"

Hands fly out, casting the traitor into outer darkness. Shoulders quiver in a crescendo, ending in a magnificent effort that seems a shrug to end all shrugs, then subsiding in a series of minatory, twitching little shrugs, as if to show that he is not yet exhausted, except as to patience.

The captain was a pleasant, short, rather hard-bitten man,

with unruly hair and a clean-shaven, neatly furrowed face. He had a passion for poker, which was very unfortunate for him when he tackled my fellow countryman, the resplendent liar. The captain would only tear himself away from the card table to have tea with us. A taciturn man, his conversation was brief but pungent.

"Dos cosas me gustan fuertes," the captain would remark every day at tea, "el té y los besos"—"I like two things, strong tea and kisses."

We had a heavy fog while in the La Plata estuary, and after we reached the sea this gave way to heavy winds. Even under water ballast the tanker rode high, and in high seas it developed a complex and unpleasant motion. Not only did it twist and squirm with the usual pitch and roll of a long, light vessel, but also the central structure developed an alarming individuality. All day and all night long it would creakingly sway like a rocking chair, independent of the two ends of the craft. I am not of those for whom the sea is endlessly thrilling. The ocean is only water, and as for waves—"plus ça change, plus c'est la même chose." Getting acquainted with fellow passengers is amusing; remaining so after a day or so is usually boring. With sea-legs once acquired, a good storm is thrilling for an hour; but it usually has the ill grace to wear out its welcome. The great advantage of sea travel is that it makes the land seem so attractive. Fog and wind delayed us, and I was not pleased by the delay.

On the fourth day out, the weather moderated and the sun shone. We were due at our destination, but still far from it. In the afternoon land did come into sight as we rounded a cape and steamed into the somewhat turgid waters of the Golfo de San Jorge—the Gulf of Saint George—that large bite taken out of the eastern coastline of the tail of South America. I welcomed the waters of my patron saint, perhaps as an omen, but more likely because here we would land.

When we awoke on Sunday, September twenty-eighth, the *Frers* was at anchor in the roadstead of Comodoro Rivadavia.

Comodoro, a particularly small spot on a few of the most recent and most detailed maps. Comodoro, metropolis of Patagonia. Under the morning sun, steep cliffs rose to a level plateau. Strips of lower land along the shore were crowded in places with oil derricks, tanks and houses. Most of the buildings were of corrugated iron, and there were no trees or anything green. It was bleak.

"Bleak, a.," says the dictionary. "Exposed to wind and weather. Syn.: bare, barren, bitter, blank, cheerless, chill, cold, cutting, desolate, dreary, exposed, hostile, raw, stormy, unfriendly, unsheltered, waste, wild, windy."

The dictionary has taken the words right out of my mouth. Patagonia is all of that, and I may as well start repeating it now: Patagonia is bleak.

Other vessels must anchor in the open sea, far from land, but for its own tankers the government has built a long steel mole. As at most Patagonian ports, there is no harbor, only an open roadstead, and the tides are tremendous, the currents strong and tricky. After the usual long wait, the *Frers* was swung around and backed tenderly nearly to the mole, where it was tied, tail-on, in a ridiculous and humiliating posture, as if straining to escape to Africa, across the way. This permits the passage of oil lines, but is very little good for anything, such as passengers, that cannot be pumped through a pipe. Some day, perhaps, vessels will be able to discharge comfortably at a dock at Comodoro, as a concrete breakwater, wharf, and ship basin were started there some years ago. One day during our stay there we saw two workmen looking hopelessly at the embryo of this structure, perhaps wondering what to do next, but reason prevailed and they decided to put it off till mañana.

Doubts as to how we were to get to land were soon dissipated. Group by group, as our turns came, we were crowded into a little wooden box suspended by steel cables. This was lifted by a crane, whirling and swinging madly through the air and then dropped with a spine-racking thump into a dirty scow oozing tar at every seam. This lighter was nosed up

to the end of the mole by a tug, and then we were again swung
dizzily upward and dropped on a plank platform, still some
distance from land. Here we waited some more—if you are
not good at waiting, never go to South America. Finally we
were hauled to the shore by a little gasoline car running on
a very narrow-gauge track hung precariously over the water.
Another scientist, who had spent years in Patagonia making
a collection, lost the greater part of it when it was tipped
off this mole and swept away by the tide. But in our landing
all went well, and we were soon sitting on land, waiting as
usual, for our baggage this time.

As a result of these difficulties in landing, freight for the
hundreds of miles from Buenos Aires costs less than the few
hundred feet of lighterage from the ship to the shore at its
destination.

So I finally arrived on Patagonian soil, one month and
twenty days after leaving New York.

We were not yet at our destination, as the government oil
tankers operate from the main fiscal oil camp, which is three
kilometers from town and hence is almost always called
Kilómetro Tres. When our baggage had finally come, the
American salesman and I hired a car to take us in to Comodoro itself. Most of the road is cut into the cliff-like face of
the great hill, the Cerro Chenque, between Kilómetro Tres
and the town, and bits of it are forever trying to slide into
the sea. This short stretch and its continuation through the
government oil field are about all that Patagonia has to offer
in the way of good roads and it is very carefully policed, as
I later found, to my sorrow. Having built a few kilometers
on which a speed above a crawl is possible with a car, the
authorities became very nervous for fear it would be used for
speeding, and neatly prevented this by cutting deep troughs
across it, cunningly calculated to wreck any car that tries
to pass at more than twenty miles per hour.

"Why," I afterwards complained to a local resident, "why
do you build one short piece of good road in the entire stretch
of Patagonia and then spoil it with this silliness?"

"But, señor, it is an idea Yanqui. Up there in America of the North all your roads are so built."

Seen after weeks in the vacant hinterland, Comodoro Rivadavia ("Comodoro" to its familiars) came to take on all the color and importance of a big city, and it is no longer easy to recall my first discouraged impression of it. The town is a triangle, cut into small squares by its streets. The long southeast side is girt by the sea, and the north side is immovably fixed by the abrupt slope of the Cerro Chenque, rising about seven hundred feet above the water, but to the west the town sprawls out in filthy outskirts among bare grey hills. There are no trees, no grass, nothing to beautify or relieve, only squat, square rows of stucco and corrugated iron buildings, with one or two concrete structures of two stories. The population is, I suppose, some three or four thousand, and there may be as many more people scattered through the surrounding oil camps.

There are numerous small shops, but commercial life centers chiefly in the two large stores which sell everything from omnibuses to peanut butter and keep an iron grip on the economic life of the region. The houses are dreary little places, mostly of corrugated iron, that most unesthetic of building materials, vastly popular in Patagonia where all lumber has to be imported from great distances. There are one or two bedraggled, vest-pocket-sized gardens hidden away in patios, but nothing really pleasant or homelike. "Home" is a mockery in Comodoro, and almost throughout Patagonia. I met at most three or four people who planned to stay there and looked upon it as their home. All the others were there madly against their wills, and with no hope but of being able to leave.

They come to make money, in oil, in sheep raising, or in commerce. They plan to spend a few years, then to go home, to the real homes of their hearts, and live in comfort on the well earned proceeds of the years barrenly passed in Patagonia. The price of wool has dropped, oil is not the bonanza they expected, and commerce is almost dead. They cannot afford to sacrifice their investments of money and time. Year follows year without bringing the right moment to cash in. A large proportion of the population of Patagonia plans to leave "next year," and has been planning so for many years now.

For amusement, Comodoro offers three movies, two of which are usually closed, numerous bars, and seven brothels, stocked with offcasts from Buenos Aires (and even there they are not too nice in these matters). For accommodation, it offers about a dozen hotels and inns, grading from the Hotel Colón down into utter nastiness.

Arturo Bruzio, the proprietor of the Colón, greeted us when we arrived in town, Bruzio, that conflicting personality, who disapproves so heartily of excess in his customers, yet is so anxious to stimulate their purchases. His is the hotel for visiting northern Argentines, foreigners, and other madmen. It charges more than the other hotels, has three bathrooms of nearly European luxury, is cleaner, and on its good days has a better, or at least less blasting, social tone. In front is the well stocked bar and its tables, and behind is the dining room, divided by a low partition, with families and the more respectable customers in one pen and hoi polloi in the other. From visit to visit we oscillated between the sheep and the goats. Still farther back are the ground floor rooms, grouped around a patio, and upstairs is a roof patio surrounded by other rooms. As is universal in hotels of this class in the Argentine, each room opens separately onto the unroofed patio. It is hardly worth while to go for a bath when it is raining; you get so wet on the way.

Established in this home from home, I went out into the wind to explore the town. The terrible Patagonian wind! It is the constant background of any Patagonian memories. It is the most fundamental fact of life there. Only the day before I arrived, it had shown its teeth in a particularly impressive way.

An airline had been established, running almost the whole length of Patagonia and linking it with the northern Argentine.

Savage places are now almost everywhere marked by such incongruities, the impingement of some modern invention on a background still primitive. In such places civilization no longer develops sanely and harmoniously. It breaks out in spots like a disease. People who never saw a book and whose ideas of sanitation would offend a chimpanzee may possess phonographs and sewing machines. Places inaccessible by any road may be quickly reached by plane. The heart of Patagonia is within one day of high civilization by the airline, when it runs.

The day before I landed, the plane had taken off for Deseado, the next town of importance to the south. Hoping to beat the wind, which usually gains vigor with the sun, the plane took off before dawn, but on this day the wind had anticipated and early reached its full force. The pilot was hardly well in the air before he realized that he could never make his destination. He worked back to the landing field, but the gale had now reached such force that attempting to land meant almost certain death or injury for himself and his passengers. He had no choice but to try for Deseado, trusting to luck to find some lull in the wind there, but as he headed the plane into the wind again he found with horror that he could make no headway against it.

Then began a duel which lasted for nearly four hours. The powerful motor and the wind were almost evenly matched. Sometimes the plane would gain a few hundred yards, then the savage wind from the distant cordillera would retaliate and drive the plane out over the open sea. For most of the time the two were so evenly balanced that the plane simply hung as if suspended in space. All this time it was never quite out of sight from the field from which it had risen.

When a plane lands at these Patagonian fields, a detachment of soldiers hooks ropes to metal rings in the wings the instant it touches ground and hangs on to these to keep the plane from blowing away. While the plane was fighting to get back over the field, watchers from the ground had sent hurriedly for men to try to hold it, if it could land at all. His gasoline supply nearly exhausted, the pilot made a landing with superb skill, the ropes were hooked on, and the passengers safely removed. Hardly were they out of the plane, when a strong gust whipped it over and the propeller, left whirling to help hold the ship, killed one soldier and horribly mangled another.

That was not a very unusual gale, or a brief storm. The wind frequently blows so for days at a time. Its brutal force and nerve-destroying continuity almost pass belief. They have given rise to many tall tales and popular legends which gild the lily and exaggerate the immense. It is probably not true that a man once inadvertently fired a rifle straight into the wind and was killed by his own bullet as it was blown back. It is somewhat more plausible, but still not well authenticated, that a Patagonian chicken was accidentally taken by the breeze from the rear and completely plucked so that its owner had to knit it a sweater. But it is true, by the witness of my own eyes, that the wind blows stones large enough to break plate glass, that wild geese trying to fly into it are often rapidly carried backward, and that it is sometimes impossible for a man to progress against it except on his hands and knees, clutching bushes for anchorage.

So on my first afternoon in this unpromising land I fought my way miserably through the dust-colored streets. Abandoning my intention of sightseeing, I disturbed the siesta of an Englishman to whom I had a letter of introduction. He was most cordial, and promised to help me get chapas for my truck. "Chapa" is one of those delightful Spanish words with such miscellaneous meanings. A chapa may be an identification disk, veneer, sheet of metal, foil, cap, leather chape (whatever that may be), rosy spot on the cheek, rouge, good sense, a game of tossing up coins, trunnion plate on a gun carriage, and, as in this case, an automobile license plate.

In police-ridden Comodoro every car must be licensed, and every driver registered. Traffic rules in the town are strict and rigidly enforced—driving to the left, as in England, was a little bothersome at first but soon became second nature. In the narrow limits of the town and nearby camps, red tape and restriction are supreme: official revenge against the vast hinterland which refuses to be regimented and subdued. These details of registry were, however, quickly mastered in the usual way, by ignoring the set routine and using influence shamelessly.

The next few days were occupied by details of supplies, plans, and personnel. For camp and field helpers, good luck sent me Justino Hernández, and medium to bad luck sent me Manuel Laurencia. Justino, about twenty years old, was a true son of Patagonia, which, at that time, he had never left. His mother was Lithuanian, one of a large family scattered through this region, and his father was Chilean by nationality and, if appearances and common knowledge do not lie, Indian by race. Justino was born at the Codo del Senguerr, which we later visited and which is one of the most remote corners of the earth, and there he had spent most of his life. He had been to school once, for three or four weeks, but this spurred him to learn to read and write after a fashion. He was a useful compound of native intelligence, self-confidence, natural courtesy, and untiring willingness. He assured me that he was "muy baqueano," that is, well acquainted with the country and full of knowledge of its stratagems and possibilities, and he was right.

Manuel was a somewhat less fortunate acquisition, a Portuguese who with some of his countrymen had a small hut on the flanks of the great pampa. His companions were content to raise sheep, but Manuel was more restless and spent such time as he could working as a cook, a teamster, or anything else that came to hand. As no one else at all qualified for the job could be found, I employed him as cook and camp man.

The racial mixture of our small party is typical of the whole of Patagonia. Except for a small reservation far back in the mountains and for an occasional itinerant individual, the aborigines are no longer in evidence. Here, as in Buenos

Aires, the police are mostly Indians of various sorts, and Indian blood flows in the veins of many of the otherwise European, or at least Europeanized, inhabitants. The true Patagonian Indians, the Tehuelches, were never numerous, and now the "Patagonian giants" are little more than a legend. They really were gigantic, but only in a moderate sense of the word. The men averaged about six feet in height, the women considerably less, heights not abnormal yet placing them among the tallest races of the earth. There are apparently authentic records of several individuals seven feet or more in height. These big Indians called themselves Tsoneca and were called Tehuelche by their neighbors and foes the Pampas Indians. Their name Patagonian Indians was acquired from the Spanish. Patagonia is the land of the Patagones, or men with big feet. Both feet and height seemed enormous to the early Spanish explorers, who were small men with disproportionately small feet, as are their descendants today. Nordic expatriates in Latin countries often have difficulty in getting shoes large enough to fit them.

Aside from the few Indians and the much greater number of half and other breeds, Patagonia is now inhabited by the more adventurous or the less successful members of almost every nation on earth. Every Latin race is largely represented, and there are also considerable numbers of Welsh, Boers, English, Slavs of all varieties, and others of every shade of color and every sort of national allegiance. There are very few North Americans; in all our wanderings we encountered only one who was actually resident in Patagonia. This one, "El Rey," was, however, worth many a lesser man.

The nights, and not infrequently the days also, were still very cold when I arrived in Comodoro. I shivered in anticipation of nights on the pampa, and decided to imitate the Indians and go prepared with a fur robe. With some shopping around, I found a Chilean Indian woman who shared a hovel on the edge of the town with a swarm of children and a large litter of small pigs. In this squalor she produced a blanket of real beauty, a *quillango* in the best Indian tradition, made of the skins of new-born guanacos, sewn together with ostrich sinews, and delicately patterned in white and russet. All the time I was in Patagonia I slept like a king rolled up in my fur robe and requiring nothing more for warmth, and in New York I still use and appreciate this one Patagonian luxury.

The first act of my American salesman friend from the *Frers* was to hire a car and driver, and order that they be available outside the Colón at all hours of the day, and especially of the night. At one A.M. he would go reeling outside and whistle loudly, and the car would materialize from somewhere in the darkness. The driver had a name, but it proved too difficult for his employer, who rechristened him Light-horse Harry.

Light-horse Harry was one of the very few people whom I met in Patagonia who seemed really to love the country. He was an inexhaustible source of enthusiasm and misinformation concerning it. Driving around the countryside of Comodoro, he would pause at every rise in the road to express his delight.

"Qué lindo es!"—"How pretty it is!" he exclaimed for the hundredth time one day, on my first excursion outside of town.

Even there in Lost Canyon, Cañón Perdido, in the springtime, the first impression was one of utter barrenness and complete dullness. But after a time, the blue sky with an occasional cloud, the distant, purple, flat-topped hills, and the brush-covered foreground did seem to unite into a significant picture. On closer examination, the apparently dull slopes proved to be sown with minute brave flowers. Even the thorny scrub was vernally decorated with tiny, greenish, waxen bells, not visible casually. The wind hummed in the brush as if it were passing through an invisible forest. Perhaps a little of Light-horse Harry's strange enthusiasm passed into me and I fell into a mood approaching quiet satisfaction. I felt hopefully that here was a land of gusty passions and masculine restraint, turning a bitter face to the world, but hiding a savage beauty for those who could penetrate to its heart.

Land of contradictions! To this day I do not know whether I love it with all my soul or hate it with all my heart. Or both.

My strange mood grew as we approached the town at sunset. The wild shore seemed to hide something essential to life, if it could only be found. The tide was out and the surf boomed beyond the wide strand. Beyond us, a great headland stood out against the sky, and beyond that another still more distant, almost lost in the purple haze. On this the orange rays of the setting sun shone dimly.

That evening we had cocktails in the Club Social, the gathering place of polite society in Comodoro and so eminently respectable that it had been almost a year since anyone was shot there. The scene was so thoroughly ordinary that it seemed madly fantastic. This pleasant room was thousands of miles from the old centers of civilization. These comfortable-looking people were pioneers in the midst of a great rocky wilderness. They danced, and drank, and gossiped, while outside the wind howled down from the pampas of Patagonia. Patagonia, whose very name is a synonym of desolation, Patagonia, where smug and unimaginative Comodoro is man's only slender foothold and where nature still makes a successful stand against his inroads.

During my first week there, talking pictures came to Comodoro. Some unusually bold entrepreneur had brought down a portable sound apparatus, to bring to the Patagonians the last blessing of modern invention. All the élite attended, tensely anticipant as the house darkened and loud rasping noises began to issue from behind the screen. A bearded face as big as a house suddenly blazed forth. Its lips moved, and after a few moments there came a thin soprano voice, rising and falling like a siren as the ungoverned machine speeded and slowed. An awesome "Ah!" arose from the packed house. But afterwards, the Rivadavians were not convinced. No, indeed, talking pictures had not come to stay. They could never replace the silent film. It was only a passing craze of the jazz-mad generation of young and effete Patagonians.

So rapid is progress that when I left Comodoro for the last time, it had three theatres wired for sound and producing very creditable effects.

The purchase of supplies for our camp presented no great problem, thanks to the Sociedad Anónima Importadora y Exportadora de la Patagonia, whose breath-taking name is senselessly abbreviated to "La Anónima," "The Nameless." The Anónima and its somewhat less extensive rival, the German firm of Lahusen y Compañía, are to Patagonia about what the Hudson's Bay Company was to Canada. Only the most powerful estancieros escape their domination, and the small landholders and sheep ranchers are often little more than minions of these great corporations. They practically monopolize the importation of supplies and the exportation of wool from Patagonia. They own great estancias, sheep ranches, themselves. They run lines of cargo and passenger boats. The important, and in many places the only, stores are theirs. Each year they provide the small holders with supplies, taking in return a mortgage on the next wool crop. In recent years, at any rate, the crop has been insufficient to pay off the debt; it is bought by the creditor companies themselves. The rancher must again and still more deeply go into debt for new supplies. And so year after year he is never out of debt to the companies, and seldom or never sees any actual cash. His only escape is to abandon everything, and then he can only leave on a company boat, if he has the fare.

The big store of the Anónima in Comodoro is the apotheosis of the old general stores of our country crossroads. It has at most a few thousand customers, and its space is not a tenth of that of an average American department store, but into that space it manages to crowd almost anything that can be bought anywhere: needles, caviar, omnibuses, rifles, ponchos, dolls, furs, and ham. Gasoline motors rub elbows with cases of whiskey, and evening gowns seem to frown on bicycle pumps. The poor herdsman in the country may have almost nothing he has not made with his own hands, but his brother in the town hardly needs to envy Buenos Aires for its shopping opportunities.

Gasoline for our truck was carried in three built-in tanks, and to increase our cruising range we also usually carried two large two-hundred-litre drums, which were left at base camps and drawn on for reserve. After long argument, it was arranged to buy gasoline at wholesale prices. At that time the retail price was twenty centavos per litre, about thirty cents a gallon, although purchased direct from the refinery and within sight of the wells that produced the crude oil.

With men hired, supplies purchased, and equipment revised and packed, it was still necessary to delay for a couple of days. Two government geologists of the fiscal oil corporation were to accompany me into the field, and they were unable to leave as soon as I was ready. To pass the time and to help orient myself in the region, I spent two days with the manager of the nearly defunct Solano Oil Company at its oil field a few leagues north of Comodoro.

The Solano Company, now with only six or seven producing wells and planning to cease operations altogether, is one of several corporations, financed abroad, that have come into the Comodoro field. There are subsidiaries of the Anglo-Persian and of the Shell interests and several private producers, but the government has seized the lion's share of the field and discourages competition. Standard Oil flirted with the field for a time, but withdrew discouraged. Things have not gone as well as was hoped, either for the government or for most of the private interests. The local field seems to have been well explored and to be near its maximum production, yet it does not begin to supply the needs of Argentina and has no importance in the world market. The production per well is small, and the petroleum is heavy and of a poor grade. Energetic and expensive prospecting has not as yet revealed any significant extension of the field or any new fields in this general region. The government's plant is very extensive, with about a thousand producing wells when I was there, a small refinery, and several thousand employees, yet I was assured that it had never shown a profit. It seemed to have the fault of bureaucracies everywhere, and particularly in Latin America, with several men to do one man's work and no real concern for profit and loss.

Like most of the companies, Solano had built a very comfortable camp, now almost uninhabited, and I spent a pleasant time there. My host was a very congenial Dutchman who had roamed widely over the earth, particularly in the East Indies, and had finally drifted into Patagonia, where he was in a fair way to becoming rooted, having acquired an Argentine wife and a responsible position.

We spent an afternoon wandering about among the hills, which, like most of those near Comodoro, are formed by the uplifted deposits of an ancient seabed. In them are innumerable shells of fossil oysters, now unfortunately extinct. If only they had survived, what a feast they would have made for Gargantua! Many of them are over a foot in diameter. In one place an ambitious partnership had built a crazy and precarious system of trails on a steep slope, along which, at very real risk to their necks, they drove an old Ford truck, collecting the fossil shells. Some millions of years after the animals that formed them had died, these shells were being converted into lime in an amateurish kiln at the foot of the slope. These noble oysters, by the way, are known to science as Hatcher's oyster, named for a distinguished American explorer of Patagonia in the nineties, John Bell Hatcher.

I awoke the next morning at Solano to find the whole country white and a heavy snow still falling. The season corresponded to early April in the Northern Hemisphere, and this was the last snowfall of the season in this warmer coastal area. It stopped snowing by noon, and by evening the constant wind had melted the snow except in sheltered spots. We skidded and splashed our way back to town over the perilously slippery roads.

CHAPTER III

OFF INTO THE WIND

From my journal:

Casa Ramos
Oct. 11.

FOR almost the first time on this expedition, hopes were justified, and we did leave Comodoro this morning. Justino and Manuel came to the hotel, and we drove to the baker's for the last of our supplies, a gunny sack full of galletas, hard, round, dry, heavy rolls or biscuits which keep indefinitely but can hardly be eaten except by breaking them up with a hammer and soaking for some time. Then we went out to Kilómetro Tres, where the government geologists were waiting for us: Piátnitzky, a soft-spoken Russian, and Conci, a diminutive, plump Italian. In geology as in most other sciences, there are few native Argentines. They took their own truck, chauffeur, and cook, and were rather shocked, I think, that my ménage does not include a chauffeur and that I drive myself. Even in the field, scientists in Latin America are not supposed to do any physical work, which perhaps has helped to retard scientific progress here.

At about nine, we left Comodoro. Into the field at last, after over two months, thousands of miles of travel, a new world, a revolution, weeks of effort, delay, and preparation!

At first we drove south from town along the hard beach. Our departure had to be set for low tide as this exit is closed at other times. At the South Camp of the government field, where they are drilling some wells right out in the sea, we turned inland and started the long climb to the top of the Pampa de Castillo, here over two thousand feet high and

38

standing as a great barrier to all intercourse between the coast and the interior.

It seems that the pampas of Patagonia are not the low plains of my school geographies, but the flat tops of large plateaus of various heights. The "plains" of Patagonia are often higher than the surrounding mountains. Of these central Patagonian pampas, the Pampa de Castillo is the largest and the highest, extending for about 150 miles roughly parallel to the coast and with its surface varying in width from a few hundred yards to thirty miles. This bulky barrier has played a leading, and often sinister, rôle in the development of the region. It is so difficult of passage that the coastal zone had been visited, at long intervals to be sure, for some three hundred years before anything was known of what lay beyond the pampa. It is one of the principal reasons why this was the last part of Patagonia to be explored and settled.

Thanks to the possibility of using motor transport, we have just crossed it easily in a day. Even now this is a dangerous trip in winter, and a number of men foolhardy enough to attempt it have lost their lives by freezing or starvation.

Our road, or rather our track, for it is little better than driving cross country, ran up a long cañadón, a narrow valley carved in the steep edge of the pampa by ages of wind and rain. After an hour or so, the cañadón became visibly smaller and less deep. Patches of snow began to appear along the way. Suddenly we emerged onto the broad, gently rolling surface of the pampa.

Just after reaching the top, we could see the now distant blue ocean down another and shorter valley, hardly a glimpse before it was gone. So we traveled over the Pampa de Castillo. Its surface has no drainage and the melting snow has filled every depression with water, so that we passed between and through a network of puddles and small lakes. The wagon track has in many places been rutted deeply and the water collects in the ruts. Much of the time we were driving in one or two feet of muddy water, but the bottom was usually firm, thanks to the pebbles.

I remember reading about these pebbles in Darwin's "Voyage of the Beagle," and I had looked forward to seeing them. They cover the whole surface of the Pampa de Castillo—in fact, as I recall from Darwin and others, they cover every flat surface in Patagonia, extending for hundreds of miles from the Straits of Magellan to the Río Colorado. As we saw them today, they lie in a thick mantle, many feet in depth, are well rounded, and are of quite varied sizes, but mostly from the size of a walnut to the size of a man's fist. Most of them are porphyry, and some of rich colors, yellows, oranges, browns, and reds.

Since Darwin first saw and speculated on these "rodados patagónicos," about a century ago, argument has raged as to whence they came, how, and why. As far as I know, the mystery is not much nearer a solution, except that Darwin appears to have been wrong in supposing them to have been deposited by the sea. Tremendously impressive in their vast extent and important in helping to make Patagonia the sterile desert that it is, they remain a Patagonian marvel and one of the earth's many mysteries.

There are many thorny scrub bushes on the pampa where we crossed it, but no grass and no trees. The general effect is one of a great emptiness. A vacancy not expansive, as on a true plain where miles of country may be seen from every little elevation, but cramping and rather unpleasant.

In the distance a dark purple peak became visible, the Pico Oneto. With little warning, we reached the far edge of the pampa, and there suddenly opened before us a really grand panorama. In the farthest distance the dark serrations of the Sierra San Bernardo and beneath us the vast Sarmiento Basin, with Lake Colhué-Huapí gleaming like silver in the sunlight. Rimming the basin and here and there in the foreground, isolated peaks, some chalky white, others of somber volcanic rock. Immediately before us dropped away the jagged, dissected edge of the pampa, bare of all vegetation. In the dry air and beneath the clear blue sky, these hundreds of square miles seemed like a boldly graven miniature model of a continent.

Contemplation was rudely ended as we entered the Cañadón Grande and jolted our way painfully down it at about five miles an hour. After three or four leagues—the Patagonian unit of distance is the league of five kilometers, a little over three miles—after three or four leagues of this cross-country reeling we came to a wildcat well that the government is drilling for exploration. It is drilling several of these forlorn hopes, endeavoring to extend the field, but so far has found no commercial prospects outside the immediate environs of Comodoro. The government geologists stopped for a few minutes to examine drill cores, then we continued down the cañadón toward the lake.

The way now lay in the midst of badlands, scenes of our future labors, until we came to a house on the water's edge, the Casa Ramos. Ramos, who is away, must be a superior character, as his house is well above the average. It is roofed with corrugated iron, but the walls are of adobe bricks and there are several rooms, each with a separate outside entrance. One of the rooms has a window, without glass. On one side of the house is a small and dejected garden, to which water is pumped from the lake—I am proudly told that this is the only garden patch for some thirty leagues around. On the east side of the house and thus sheltered by the house itself from the constant west wind, are two small fruit trees now in blossom. All the doors also are on this leeward side of the house. Nearby is the well, only a few yards from the lake but with a lower water level. The lake water is fresh but muddy, the well water clear but slightly saline. Take your choice. Back of the house are corrals, laid out with living thorn scrub, now with yellow blossoms, supplemented by piled-up brush. On another side is the conical outdoor oven of mud, like those of New Mexico. Indeed the whole scene reminds me more of New Mexico than of anything else I have seen before.

A table, the only one in the place but another evidence of refinement, was brought out and set beside the fruit tree; they are too small to set a table under them. Here the aristocracy, that is, Piátnitzky, Conci, and I, ate old sausage and bread, while the rabble ate other things and elsewhere, the first glimpse of unexpected social distinctions here where, if anywhere, one would look for complete democracy. It was sunset and we had not eaten all day.

After this refreshment, the three of us strolled along the strand talking. Here at Casa Ramos tonight there are two Argentinos, a Spaniard, an Italian, a Russian, a Portuguese, a North American, and a lad half Lithuanian and half Chilean Indian. Among us we can talk eight or nine languages, but no two of us can converse in any language but Spanish, which half of us do not speak at all fluently. At any rate, Russian, Italian, and North American, we strolled about conversing in stilted Spanish. I have tried to savor this evening to the full, for I am warned ominously that I need not expect ever to see another like it down here. There was not a cloud in the sky, and, as if by a miracle especially arranged for my arrival, not a breath of air stirring. The laguna—for we are not on the main lake here, but a small and shallow arm of it,—the laguna was calm and like a mirror, faithfully reflecting the surrounding cliffs and peaks and the orange and purple of a desert sunset. I expected much of Patagonia, but I did not expect it to be beautiful.

After dark we had some soup and puchero, and then went into the house and sat for a time with Don Francisco, who seems to be acting as host in the absence of Ramos. The main room has a dirt floor and whitewashed mud walls. The furniture consists of a cupboard, a small table, and some low benches, more like footstools than seats. In one corner a fire is built on the floor, some of the smoke finding its way out through a baked mud flue extending out some distance above it. Aside from the glowing coals of the fire, the only light is given by a small bottle of oil, hung on the wall with a burning piece of rag dipped into it. This homely imitation of the old Greek and Roman lamps gives considerably less light than a good wax candle.

We all sit in a semicircle around the fire drinking maté. Fortunately I learned some time ago to tolerate and even to enjoy this Patagonian cup that cheers but not inebriates. Don Francisco sits next to the fire, his picaresque face gleaming in the dull glow. He is serving the maté, testing the scalding water from time to time by pouring a few drops on his index finger, passing the gourd to us each in turn. For a long time no one speaks. Finally Don Francisco sighs.

"Es triste, no?" Sad, isn't it?

All sit wrapped in profound thought for another quarter hour.

"Sí, sí. Triste," finally volunteers a mournful voice in the shadows. All sigh deeply, and relapse into permanent silence.

Camp No. 1
Lake Colhué-Huapí
Oct. 12.

We arose this morning at about 5:30, shivering in the bitterly cold dawn. Manuel soon had some hot coffee for us, and we started off, at first over the usual wagon track, and then on no road at all, which was rather better going. These trails and tracks are useful only as guides. They certainly do not make travel any easier or speedier. In some places just by the lake there are broad patches flooded in the wet season and when dry these make perfect speedways except that they end suddenly and dangerously in steep-cut banks. There are local patches of sand, but no considerable development of real sand dunes. On the higher flats driving is made difficult because the wind carries away the sand from the open places, leaving them floored by pure cobbles, and piles the sand up around every bush and shrub. Instead of being soft, like our western sagebrush, and offering little hindrance to a car, each bush thus comes to have a core of packed sand which makes travel over it very slow and painful.

Piátnitzky and Conci had promised to show me a strip of

badlands which contains fossils and will be a good point of attack for my work. We soon were in sight of the badlands, or rather of a great barranca, cliff, that forms the southern rim of the lake basin. Near the foot of the cliff we abandoned the cars and began to walk. We passed some ostrich eggs, last year's and empty. It is early for this year's eggs. Farther along, four guanacos started up and disappeared into the badlands. They will have young soon. These creatures are still new and strange to me.

Don Francisco, who came along with us, urged us on and up the high cliff, promising us that there were fossil bones near the top. "A cemetery, señor, a true cemetery," he cried encouragingly when our enthusiasm seemed to flag. I was not very sanguine, having followed many a local guide elsewhere in search of fossil bones that he claimed to have discovered, only to find that they were recent bones, or not bones at all. But Don Francisco's "cemetery" was the real thing. On a ledge near the top of the barranca many fragments of petrified bones had weathered out and were scattered on the surface. The only thing we saw that was much good was a large tusk, curved like a saber and two or three feet long, which I took. Certainly other and better specimens can be found there with careful searching. The prospect warrants camping here and starting intensive work.

We returned to Casa Ramos for lunch, loaded up the truck again, and drove to a camp site several leagues away. It is right on the lake shore and at a little adobe hut which is occasionally used by passing sheepherders but not regularly occupied. The afternoon was consumed in putting the camp in order. For the present we are using just two tents, one for me and a larger one for Justino and Manuel and also as a kitchen and dining room. They can also use the hut if they like. I am curious to see how the tents stand up in the wind. What an argument we had with the tentmaker in New York, getting him to use canvas of double strength and to put on double guy ropes! He was sure that the tents would be so stiff and heavy as to be useless. They are certainly not too strong, and my only hope is that they are strong enough to stand the wind for a season. The pegs are long pieces of angle-iron, with rings welded into the end to hold the ropes. They should hold well even in this rather soft soil, which is really nothing more than a flood beach of the lake.

We were fortunate this afternoon in finding a sheepherder temporarily staying at another hut a few kilometers away. He had just butchered a sheep, and we bought half of it from him.

As I ate my soup and puchero this evening, Justino and Manuel looked on silently like hungry wolves but refused to touch a bite until I was through. I gulped my dinner down, feeling rather embarrassed and very much isolated. About all I have in common with my companions is that we are all human souls. That is a surprising lot, and enables us to get along well together, but we speak different languages, are of different races, have spent our lives in different hemispheres of the earth, and have backgrounds and habits of thought which are almost absurdly divergent. Yet here we are.

So early to bed.

Oct. 13.

And early up. It was drizzling, but after coffee and galletas Justino and I went off to the barranca, driving the few kilometers to a point below Don Francisco's cemetery and then climbing up. I was much encouraged by finding some good prospects, but everything was so wet that we could not do any work on them. It began to rain hard, so we gave it up and returned to camp by lunch time.

It rained hard all afternoon and we stayed in camp. I read a book on the geology of this region by the famous Argentine scientist Florentino Ameghino. Manuel and Justino spent most of their time doing nothing, a really difficult art, almost impossible for a North American to acquire but second nature to the natives of these parts. Conversation does not flourish in our group. My Spanish is improving rapidly, I think, but is not yet really fluent. Even aside from linguistic difficulties, it is impossible to converse with people who simply agree vehemently with everything one says, as they do with me. Perhaps this is shyness, rather than exaggerated respect, but it does seem to embarrass them if I attempt to do anything for myself and their politeness is almost painful. Justino, particularly, waits on me hand and foot, even tucking me in bed at night. But they are good lads.

Oct. 14.

Not one of our better days!

The morning was vile, a nasty drizzle, clouds hiding the top of the barranca, cold biting in to the bone. It was hopeless to try to dig fossils, so I decided to go to the nearest settlement to get acquainted and to try to get such odds and ends as inevitably prove to be missing when camp is first set up. The truck refused to go, and we worked all morning on it, wet through and blue with cold. Finally after lunch it went, although badly, and we went off to 103, a stop on the railroad about eight or nine leagues from here. Just why there should be a railroad in this region is not apparent at first sight. It runs from Comodoro inland, westward roughly, to the small settlement of Colonia Sarmiento, and was built by the government fifteen or twenty years ago. Of course there is no north or south rail connection and the only real use of the line is to shorten the wagon haul for wool bound to the coast where it is shipped north by boat. The railroad does not begin to repay the great expense of building it, and my bewilderment as to why it was ever constructed is only increased by the frequent statement that it was built regardless of profit for "military reasons." My inferiority to Argentine officialdom is increased by the fact that they can see, and I certainly cannot, what military purpose is served by a short and barely passable, isolated single-track railway running from the coast into the dead center of Patagonia, far from any national boundary or any mineral or other important natural resources.

103, so called because it is at the one hundred third kilometer post of this line, is a wool-loading station on the Pampa de Castillo. In the good old days when wool paid a profit, estancieros from the mountains used to haul by wagon to this point and there transfer to the railway. To help them to spend their profits, the Anónima built a branch store there, and now that there are no profits to spend, the store remains there, apparently from sheer inertia. After some searching I found a lad who grudgingly unlocked the store. He could say three words of English and proudly used them, although he did not know what they meant. When he understood my purpose, his face lit up. Here at last was a chance to get rid of some of the stock left over from the fine, freehanded days when the store was built! He offered to sell me champagne, caviar, silver horse fittings, silk handkerchiefs. When I refused these and other luxuries and made my modest request for daily necessities, he was terribly crestfallen.

"Oh, señor, we do not have those. But just see these fine gold watches!"

We could not do business, and his sorrow at seeing his customer for the year depart was almost hysterical.

The truck still went very badly as we drove back. We went down Cañadón Pedro, a narrow pass down the steep scarp of the pampa, and across the Valle Hermoso. Valle Hermoso—Handsome Valley—is neither handsome nor a valley. It is a very barren, almost perfectly flat, waterless plateau flanking the higher Pampa de Castillo. The clouds had cleared somewhat and there was a gorgeous sunset. Night fell before we reached the far side of the Valle Hermoso and bumped our way down into the basin. It was long after dark before we finally limped into camp, half frozen, exhausted and starved.

Oct. 15.

The truck refused to go at all this morning. Justino was confident that he could fix it. I did not share his confidence, but left him working on it and walked over to the barranca. I had intended to work on the things Justino and I found the other day, but on the way there struck a richly fossiliferous spot and worked there happily and successfully all day.

While I worked, a guanaco appeared across a narrow gully from me and watched me curiously, yammering at me from time to time. They are all adults at present. Manuel tells me that the young are born on November tenth—he is very definite about the day. We have seen some every day, but so far not more than eight or ten together. Ostriches are fairly common. I saw six in a group yesterday, but most of them seem to be solitary or in pairs. They haven't begun to lay, although there are many dud eggs lying around from last year.

I climbed a very high cone in the badlands. Part way up the precipitous slope was an ostrich skeleton—serves the bird right for trying to climb mountains when he can't fly. On top were two beautiful eagles which have a perch there to look over the country for prey. They flew about swearing at me in raucous voices. Later I saw one of them chased by a small bird which was teasing it, repeatedly flying at its head. The eagle screamed and grabbed for the pest, but the midget evaded it easily and only renewed the attack, apparently from pure deviltry.

So back to camp. Justino failed to get the truck going and had set off on foot to try to find a man about twenty miles from here who is supposed to be a good mechanic. He should be back tomorrow. Coley is supposed to be the mechanic of this outfit, and if he isn't able to rejoin us soon I shall have to engage one here, if possible.

Oct. 16.

Manuel is extremely depressed because we have no garlic. It is one of the necessities that the Anónima at 103 does not sell. Manuel has pepper, chile, tomato sauce, mustard, onions, comino (I wonder what this spice is in English), laurel, and several other condiments to season our daily mutton, but these are but vain delusions if we have not garlic. He is also worried about meat because we have only one side of mutton left.

Justino and the mechanic arrived at about 10:30. The mechanic broke a spring on his own car getting here. Trails in Patagonia are marked by Ford springs, like camel bones along the old caravan routes. He shortly had the car going again and we went off to our daily, or what I hope after these delays may become our daily, stint at the barranca. I sent Justino off for meat, but he only succeeded in getting the hindquarters of a wether. Anyway that will hold us until tomorrow. We apparently must count on getting two sheep a week for the three of us, as appetites are hearty here and mutton is our chief and almost our only food. At about 5:30 it began to rain again, so we went back to camp.

Every day the view out over the lakes from the top of the high barranca is different and it seems an unending delight. This afternoon, cloud shadows swept across the great basin and shafts of light would pick out first one then another of the distant peaks, transforming them from a dark velvety purple to a vivid blue. Tonight the sunset was a small crimson conflagration on the Sierra San Bernardo instead of the usual garish sky-wide display.

The wind has shifted and the water is eight inches lower on our beach. Since that rare, windless first evening, there has been a steady gale and every night I go to sleep by the sound of the heavy surf breaking on the shore a few steps away.

Oct. 18.

Yesterday was a day of work, devoid of especial interest except that we found a fresh ostrich egg, the first of the season, and had a magnificent tortilla for dinner.

Today Justino and I also spent the whole day on the barranca. The weather was very ugly in the morning, with a bitter wind and intermittent hail storms. As I hung desperately fifty feet up an almost vertical cliff, the wind pelting me at forty to fifty miles per hour and alternately laden with hail and with sand and pebbles, while I tried to chisel out a fossil jaw from the reluctant rock, it occurred to me to wonder why I did not take up copra trading in the South Seas.

The cloud and light effects, however, atoned by being even more brilliant than usual as small storms passed across the basin one after another. At lunch time we found a more sheltered ledge on the cliff and Justino found some sticks and heated water for maté. We must have been a strange sight huddled there sucking our matés. It even struck Justino, whose sense of humor more often runs to slapstick, and he burst out laughing about the life of the fossil hunter.

Coming home after sunset the car refused to march at all, so we abandoned it and walked about five miles into camp. There was no moon, but most of the sky was clear and the stars were brilliant. The walk was exhilarating, but bed will feel unusually good. A hard day.

Colonia Sarmiento
Oct. 19.

Justino, prince of field assistants, drove into camp with the truck just as I was getting up. He got up before dawn, walked out, fixed the car, and drove back in triumph.

In a few days I want to go to Comodoro to see whether Coley has arrived, or any word from him, and before then I want to make something of a reconnaissance through the country that Justino knows best around here. He is only loaned to us by his uncle, for whom he works, and I am afraid that he may have to leave us. We therefore got a little mutton under our belts against emergencies and started this morning for the town of Sarmiento.

The Sarmiento Basin, Cuenca de Sarmiento, is very little above the level of the lakes which occupy so much of its area, and the part across which we had to travel to reach Sarmiento is crossed by many streams and sloughs carrying water from the basin's inlet, the Río Senguerr, or from the higher lake, Lago Musters, to Lago Colhué-Huapí. There are no bridges here, in fact no formal roads at all, and the water is high now, so we had some difficulty getting along. We got thoroughly stuck in two sloughs and had to dig our way out, and this and the general hard pulling worked the car so hard that we almost ran out of gas.

Just as we were beginning to lose hope, we saw a very ramshackle Ford parked on the open plain and made for it. In its shade were squatted several sheepherders, including a female of sorts, all as dilapidated as their vehicle. Two of them were sound asleep and the others were passing the time drinking crude Mendoza wine out of a goatskin bag. They politely passed the bag to us, and then generously siphoned a few liters of gasoline from their tank and sent us rejoicing on our way.

The tracks into the village are under deep water, into which we plunged, having no choice, and we were soon at a complete standstill, bogged down well above all four hubs. The motor, which probably had imbibed too much muddy water, also gasped and quit dead. Justino climbed out on the fenders and wiped out the motor, while I plunged into the sea of ooze and dug and inserted such brush as I could collect. We were within sight of the town, and so were soon surrounded by a crowd of urchins who came riding up on horses, bareback, with three or four urchins per horse. They were very pleased with our predicament and their comments were more derisive than helpful. A team and wagon also passed disdainfully—it was driven by a grim-visaged señora in black who was apparently taking her companion, a heavily veiled young lady, home from church. Finally we got out of the mud and gasped our way into town. There we put the car in order and filled up the tanks, but our little mishaps had so delayed us that it was too late to go on. We were disappointed if we expected the proprietor of the leading hostelry to be elated at this windfall.

"And why not?" was his only comment. "There are rooms. There are beds."

He did not exaggerate, for we are his only guests.

I think Sarmiento would be the most desolate town I have ever seen except for the mixed blessing of having an abundance of water. It is situated between the two great central

Patagonian lakes and only a few miles from either. In the
spring, as now, they so exceed their shallow basins that the
town itself is almost a lake. This impedes traffic, but it does
promote vegetation as does also the silt which here forms
the soil instead of the otherwise ubiquitous pebbles. The
townspeople are engaged in setting out trees and are even cre-
ating a central plaza, universal in north Argentine towns but
almost unknown in Patagonia. All the trees are poplars because they are the only ones that will stand up in the wind,
and even they lean strongly eastward and have all their
branches on that side. They are only saplings, but in a decade
or so Sarmiento may be a real oasis if the trees live.

I should guess the population to be about four hundred,
and most of the inhabitants keep hotels. Unless a primary
school and a church can be so classed, there are no amusements.
Creature comforts, such as running water or heat, are quite
unknown, but there is a small electric plant which dimly
lights the common rooms, but not the bedrooms, of the more
ambitious hotels and inns.

As it is Sunday, everything is closed, but we ran onto the
village stationer on the street and he opened his shop and
sold us a couple of magazines, one three and the other five
months old, which we took back to the hotel and read as we
sucked our evening maté. All the town's leading lights had
gathered in our hotel's common room, which is lobby, restau-
rant, lounge, and bar rolled into one. They were soon engaged
in violent political discussion, which Justino watched warily.
He was just trying to ease me out through a door, assuring
me that the heated politics and bad cognac were going to
culminate in shooting in another moment, when the proprietor
came hurrying in followed by his entire male staff. They
seized the most vociferous disputant and literally threw him
into a mud puddle in the street. There were immediately
loud shrieks and unladylike words in the ladies' parlor next
to us, issuing from the wife of the ejected politician. The
proprietor's wife then appeared at the head of the entire female staff and threw the lady out to join her husband in the

puddle, and quiet descended on the Rojo Hotel once more. Solicitous Justino, who had been fussing around like a hen with one chick, trying to keep me out of the line of fire, heaved a sigh of relief and went off to bed.

After he withdrew, the probable reason for Justino's excessive concern was explained when one of the loungers told me that in one of these disputes Justino himself had been so overcome with a righteous cause that he had knifed and barely missed killing a member of the opposing party. Justino has already told me, rather shamefacedly, that he doesn't think he had better drink while he is with me, because when he does he always takes too much and his head turns and he does "stupid things." As he would sooner go about naked than without his knife in his sash and his revolver on his hip, the lad may bear a little watching himself.

This hotel is built on a corner in the form of an L, with the domestic establishment of Señor Rojo in one wing and the guest rooms in the other, the common room occupying the corner. My room apparently is the bridal chamber, for it has the luxury of a window opening onto the street as well as the door which gives on the open courtyard, as usual. The window opens onto the street only in a manner of speaking, however, for the shutters and the window are both tightly nailed shut. I shall have to sleep *à la criollo*, completely without air, as the open door would expose me to the noise and the scrutiny of everyone who has occasion to use the outhouse in the middle of the patio. (Chic Sales would probably be interested to hear that this necessary adjunct, although otherwise built in the manner of such things the world over, has polished marble interior fittings—gaudy but not without certain drawbacks.)

Camp One
Oct. 20.

I slept well in spite of difficulties, and felt very sorry for myself when the lad called me at five. They seem to sleep late in the effete town, for no one was about. Justino had started a fire in the kitchen, so we went there to try to get warm and to drink maté before starting again on our journey.

We were bound for the Codo (or Vuelta) del Senguerr, the point where the Río Senguerr turns around the southern end of the Sierra San Bernardo and starts northward toward Lago Musters. The direct route, along the river itself, is admitted to be impassable, and when these people say a thing is impassable, it *is*, so we took a longer route, detouring to the east and making a swing across one of the high pampas. We bumped along in low gear, dragging through deep sand and mud and several times fording watercourses so deep that they flooded the engine.

"A few years ago," said Justino proudly as we stopped to let the motor cool a little, "this road was bad, but we got tired of it, so we all got together and laid out this beautiful new track!"

Again, when we were stuck for half an hour in a slough:

"Just see how easily we can get through now, with a little digging! Before we laid out this fine new track, I was hauling wool with a Ford truck and we were stuck three days in that slough!"

Finally we got out of the basin and climbed up the steep side of the Pampa María Santísima. This Most Holy Mary Pampa is a plateau almost as high as the Pampa de Castillo, but much smaller, separated from the big pampa by the Valle Hermoso. Part way up, we left the track and drove across country for a league or so to see a fossil forest. It was very impressive, much like the petrified forests in Arizona, although of less extent. It seems even more imposing, however, in this country now absolutely devoid of native trees. A petrified stump several feet in diameter standing among the stunted scrub thorns of today gives a very vivid idea of the change of climate in the course of geological ages.

Just before reaching the top of the pampa we turned aside again to visit an uncle of Justino's who has a hut here, tucked away in a hollow, out of the full force of the wind. The uncle and several other men were at home, and of course we had to stop for the inevitable round of maté. As it was lunch time, politeness also demanded that we stay and eat. The meal consisted of boiled sheep's head and soggy fried bread. They were much upset because they could not offer me the sheep's eyes—the head had been on hand for some time and these delicacies, which should by rights have gone to me as the guest of honor, had already been consumed. I successfully concealed my disappointment, and a hearty appetite enabled me to stuff down meat from the jaw which had been knocking around for so long that it appealed even more strongly, if less pleasantly, to the nose than to the palate. The Patagonian taste in meat is much less particular than ours. We insist that our meat shall not be fresh, nor yet so spoiled as to be obviously passé to the olfactory sense. They, however, will eat meat fresh from the animal, before it has acquired the tenderness and savour to which our butchers have accustomed us, and will also eat and relish it long after it would seem to us unfit for anything but burial.

Thence we drove over the top of the pampa for a few leagues, then down the other side into the Valle Hermoso, along it for a few leagues more, then finally down the precipitous side of the valley of the Río Senguerr and onto the narrow strip of bottom-land at the Vuelta. Here is the Casa Hernández, home of Justino's family. He had not been home for eight months, and we were soon in the midst of a crowd of dogs and people. There are eight children, but only two were at home. Hernández madre is a vast, patient, amiable Lithuanian from whom speech proceeds in a drawled cadence, punctuated by frequent exclamations of "Mirá!"—a sort of "Well, only fancy that now!" She is the only person I ever heard drawl Spanish, which, in spite of its marked rhythm, is a staccato language. Hernández padre is a Chileno, a chubby little man with an immense mop of black hair which falls into his eyes and makes him look like a jolly Eskimo or a hirsute Teddy bear. If the gossip I heard in Sarmiento yesterday is true, he is not as harmless as he looks. He has a great scar on his neck, and I was told that this was acquired at a party at his house. He became bored with one of his guests, and with that directness which seems to be so charmingly typical of the Patagonian character, he drew his knife and endeavored to kill the guest. The latter's friends took umbrage at this attention on the part of the host, and one of them snatched up a shotgun that was lying nearby and tried to shoot off the head of Hernández padre. His hand was unsteady and he only succeeded in shooting away half his neck, a wound which would have killed most men forthwith but which only temporarily inconvenienced the tough little Chileno.

The Casa Hernández is distinctly the best house that I have seen outside Comodoro, even better than the Casa Ramos. It has dirt floors and mud walls, of course, but the walls are whitewashed, the rooms are large, numerous, and well lighted, and all is scrupulously clean. There is even a parlor, which contains a table, an old phonograph, and a chair. Its walls are decorated with enormous murals, naïvely executed by the children in clashing primary colors. The masterpiece is a view of a fort being attacked by a fleet of battleships.

Unfortunately we were pressed for time and could stay only a few minutes, but promised to come again at the first opportunity. Going back to camp we made amazingly good time, averaging nearly fifteen miles an hour. Much of the way was across the flats, where it is often a few hundred yards between severe dips and bumps, we did not get stuck, and we were aided by a following wind, as we were bearing eastward.

My lads assure me that I have not yet seen it at its worst, but already I can see that the Patagonian wind deserves its reputation. Today it was impossible to drive in top gear when facing into the wind, and driving across the wind it required most strenuous efforts to keep the car from being blown off its course. The whole plain was one vast cloud of sand and dust, with only an occasional glimpse of the tops of the surrounding mountains. Most of the time it was like driving in a dense fog. It was amusing and impressive to watch the abutardas, large, strong-flying birds something like our wild geese, trying to fly in the wind. They couldn't go westward at all, at best rising or falling vertically despite their strongest efforts, and more often actually being carried backward.

The weather continues very cold and disagreeable, even aside from the wind. It is really too early in the season to work efficiently. If it were not necessary to accomplish the greatest possible amount in a single field season, it would be wiser to begin work not earlier than November, but as it is, we are accomplishing something and are that much ahead. We already have some good fossils and are well enough oriented to go ahead efficiently when, or if, the weather improves.

When we reached camp, we found that the cook had gotten some fish from the lake, and it made a pleasant break in our mutton diet. After dinner, Justino played the guitar and sang, or shouted, country songs. He knows no formal music and plays purely by ear and inspiration, but with remarkable verve. He plays both melody and accompaniment at the same time, and does things to a guitar which I am sure must be very unorthodox but which produce remarkably varied effects and almost turn the instrument into a whole orchestra. Besides strumming chords as accompaniment and picking out the tune, he beats the strings with the palm of his hand, twists two or more together and lets them snap apart, drums on the sounding board, and in general makes the guitar do almost everything but say the words.

Tomorrow I think we will work on the barranca, and day after tomorrow go to Comodoro for supplies and news of Coley.

CHAPTER IV

ANCIENT BEASTS

"The señor doctor is crazy, isn't he?" asked Manuel, not realizing that I could not avoid overhearing him.

"I don't think so. Why?" put in loyal Justino.

"He came down to this desert for no reason! No one watches him and he still works hard! And what sort of work is that? Climbing around the barranca and getting all tired out, just to pick up scraps of rock. As if there were not rocks everywhere, even in that America of the North!"

"But those are not rocks. They are bones."

"Then why doesn't he stay in Buenos Aires and get bones from a slaughter-house, if anyone is fool enough to pay him for them?"

"They are not common bones. They are not sheep or guanaco. They are very old, so antique that they have turned to stone. They are bones of animals that now do not exist, and they are not found anywhere but here at Colhué-Huapí."

"The señor doctor is fooling you. Even if they were old bones they would not be any good. What would you do with them?"

"He says that some of them are the ancestors of our own animals and he wants to learn where they came from and what they were like millions of years ago. And others are strange beasts that are not like those of today. He will put them in a museum and people will come to see them because they are so queer."

"Oh, well, then maybe it does make sense. If many of those North American millionaires will pay to see them, then he can sell them at big prices. But then he should pay us

58

more, because we are helping him and he is making his fortune."

"He says that no one pays to see the bones. And he does not sell them. He is just paid wages like us to come and find them."

"Why, then, with no one to watch him, he could stay in Buenos Aires and then go back to Nueva York and say he could not find any bones. No, friend Justino, it doesn't make sense. If the señor doctor is not crazy, he is too clever for you, and you are swallowing his lies."

And so the argument ended, as they always did, with each thoroughly convinced that he had been right all along. Manuel, like almost all the local inhabitants, was never able to see any logical reason or excuse for our activities. Rare Justino, on the contrary, took naturally to bone digging. He had tireless energy and a keen eye, as so many fossils now in New York testify. He also had that consuming curiosity which is the real reason for any scientific research, whatever higher motives we scientists sometimes like to claim for ourselves.

Yes, curiosity was the real reason why we were in Patagonia, and why others like us have been going there off and on for a century now. Indeed, there are only two valid reasons for ever visiting that bitter land, curiosity and desire for gain. Desire for gain may be more practical, but at present it is much less likely to be gratified and in the long run it is a good deal less important anyway. Those miserable flocks of sheep and that trickle of petroleum have no real significance. If they were to cease completely, the rest of the world would hardly know it, and human thought and progress would not falter in their stride. Patagonia's wealth of fossils is its real and essential contribution to the world. These fossils add chapters to human knowledge which are preserved nowhere else. And such knowledge has no ill application but can only enrich experience and give background and substance for intellectual advance. In the last analysis, most of what we are that beasts are not is due to this impractical but orderly curiosity which is called pure science.

The idea that we are the heirs of long ages of change and progress, and some real knowledge of the human rôle in the history of the earth, are relatively recent. Except for some fantastic guesses and a few neglected glimpses, this broader view of man's place in the universe has grown up altogether in the last hundred years.

By now everyone knows that life on the earth has not always been what it is today. In the mists of time loom antique monsters: scaly dinosaurs and hairy mammoths. Rhinoceroses once browsed in the lowlands of Florida. Diplodoci wallowed in Wyoming marshes. Mastodons blundered through primeval New England forests. Great marine reptiles once swam in seas where now the Himalayas stand five miles high, and strangely tusked beasts once dwelt in mountain basins now deeply buried below our present ocean level.

This prehistoric past is not a grab-bag of jumbled monstrosities. It is an orderly story as sequential as the history of human nations. Washington did not talk to Tutankhamen; and cavemen never saw living dinosaurs. The origin, spread, and disappearance of animal dynasties, their migrations, conquests, and failures, took place as definitely as any other part of history.

They now say that the earth, as a separate sphere with a hard surface, is a sprightly young planet aged at least 1,500, 000,000 years. Starting with nothing much but a lot of shifting chemicals, it gave rise in the course of time to such things as *Brontosaurus* and marathon dancers, tree ferns and Pekingese dogs, mammoths and garter snakes, each in its own time and place.

The first hint that Patagonia can shed light on some of these ancient mysteries was due to Charles Darwin. In 1831, when he was a sporting and undistinguished student of theology, Darwin set out to circumnavigate the globe in a tiny ship, the *Beagle*. He was seasick most of the time; such disciplines prepare great minds. Those five years sobered him, and he returned a scientist who was to be a chief agent in revolutionizing human thought. Patagonia played a part in the development of Darwin's mentality, and he was a protagonist in the scientific development of Patagonia. Late in 1833 he landed on these shores and, ever alert for anything strange or new, he discovered here a few petrified bones. His shipmates grumbled at his bringing these on board the ship and ridiculed the ardent young naturalist. Fitzroy, the commander of the *Beagle*, did admit years later that the bones had been of some interest. This was handsome of Fitzroy, for that somewhat testy gentleman heartily disapproved of Darwin and even wrote a curious book, now forgotten by all but a few enthusiasts like me, in which he attempted to warn young minds against following the sinful teachings of the *Beagle's* naturalist.

The bones were taken back to England and there the great anatomist Richard Owen (he also later came to view Darwin with a somewhat jaundiced and distinctly suspicious eye) studied them. Owen recognized these fossils as the remains of extinct animals, weirdly unorthodox in structure, and quite unlike any found elsewhere on the earth. His work on them was the first publication suggesting the importance and interest of Patagonia's ancient life.

It seems cavalier to pass so briefly over three of the most interesting men who ever lived: Darwin, Fitzroy, and Owen. All were great and brilliant men, as different in viewpoint and character as three men well could be, and yet strangely linked together by those few bones from such a far corner of the earth. However, I must avoid such a temptation to digress and will skip half a century of sporadic exploration in Patagonia. This is not a history, but an excuse, an excuse for my own presence in Patagonia.

It was in the eighties that the Ameghinos began their work, which was so extensive and so important that it is impossible to think of the prehistoric animals of South America without thinking of these two brothers: Florentino Ameghino, self-made savant; Carlos Ameghino, hardy and shrewd explorer.

Not long ago in Buenos Aires I attended services on the seventy-seventh anniversary of the birth of Florentino. The pupils of the Florentino Ameghino School laid a wreath before his marble effigy and sang a hymn chorusing "Gloria, Gloria a Ameghino!" They also recited in unison a poem entitled "The Wise Man of Monte Hermoso," picturing Ameghino as toiling up the sides of the mountain and on its peak, aloof from common mortals, envisaging glorious new conceptions of science, all undismayed by the fact that Monte Hermoso is not a mountain and that Ameghino's work there has been pretty thoroughly discredited. An orator then mounted the rostrum and declaimed an impassioned eulogy in which Florentino was compared with Aristotle, Newton, Darwin, San Martín (the George Washington of the Argentine), Einstein, and many others, to the distinct disadvantage of these worthies.

How encouraging that is to a poor bone-digger! One of us has schools, parks, streets, and towns named after him! The rest of us may be as obscure as last month's murderer or last year's Atlantic flyer, but our colleague Ameghino has already been dead these many years and yet is not only remembered but worshipped with a fervor almost religious. There must be a catch in it somewhere! There is, and this is it: Florentino is dead. Carlos, who did most of the hardest and most valuable work, is still alive and is even more obscure than most of the rest of us, so far as popular recognition goes. He was there at the celebration. No one seemed to mind, or even notice. There are no songs about him. "Glory to Ameghino" means glory to Florentino. Carlos spends his time waiting for the 300,000 pesos that the government once promised to pay him for the collection that he made.

Florentino did start it all. A son of poor Italian immigrants, he became a country school teacher. Walking on the pampa one day he found some bones, not very ancient, a few thousand years old, but still of a peculiar and extinct beast. He looked up the little that had been written on the subject, became enthusiastic, and decided to devote his life to this study. After working awhile in and around the Province of Buenos Aires, he became more ambitious and decided to collect in Patagonia, where it was already known that bones more ancient and even more strange occurred. He had no money, so he opened a stationery store in La Plata. The store came to be known as "El Gliptodonte"—glyptodonts, big armadillo-like armored brutes, animated tanks, were among Ameghino's favorite fossils. It was agreed that Florentino would stay home and try to earn a living, while his younger brother Carlos went off and collected bones in Patagonia.

Don Carlos spent twenty years in Patagonia, and if that land is unpleasant now, it was awful then. He wandered around, usually alone, on horseback or afoot, eating when and what he could or doing without, always collecting and studying. It wrecked his health, but he made one of the world's great collections of fossil mammals. Carlos published very little, leaving most of this to his more fluent and more sedentary older brother, but the real unraveling of the broad outlines of Patagonian geology was done by him. Of course he made some mistakes for the rest of us to correct, and left many gaps for us to fill in, but I should be very proud if I felt that I could have done as well.

Florentino named and described the new animals that Carlos found. I have not counted, but I imagine that Florentino named more different kinds of animals than any other man. He named so many that he ran out of simple names, which even scientists prefer, contrary to their popular reputation, and perpetrated such tongue-twisters as *Propalaeohoplophorus* and *Asmithwoodwardia* (after the famous English paleontologist, Sir Arthur Smith Woodward). As another wreath for his effigy and as revenge for his own tendencies in nomenclature one of the new animals found by us has been named *Florentinoameghinia*. He made known a whole new world. I am quite willing to join in the chorus of "Gloria, Gloria a Ameghino," but I do so with a mental reservation, because (and do not repeat this to my Argentine friends!) he had one very curious and sometimes very annoying failing. During his lifetime even his own countrymen viewed his work with apathy if not with hostility, and he was

almost isolated from other students of the same subject. He
became intensely individualistic and nationalistic. He thought
that he had discovered the ancestors of all the animals of the
world and of man. He thought that all the beasts and all
of humankind had originated in the Argentine. Suggestions
to the contrary only infuriated him, and when the facts
opposed his idea of things (and they do show that he was
very mistaken on this score) he ignored them or violently
fitted them into his theories somehow.

Florentino finally became director of the National Museum.
When he died, his body lay in state and the leading citizens
of the republic paid tribute to him. Carlos still lives and I
have spent many an hour drinking maté with him and talking
over the old days. He still hopes to collect his 300,000 pesos.

There had been three American expeditions to Patagonia
before ours. Back in the late nineties, while the Ameghinos
were still at the height of their activity, John Bell Hatcher
spent several seasons there, mostly in the Territory of Santa
Cruz. His magnificent collection is at Princeton University
and together with the collection in our own Museum made
by Barnum Brown, who was with Hatcher for one season, it
is the basis for a long series of ponderous and important reports, the last of which was published in 1932. Hatcher
died not long after his return, another victim to Patagonia.
In 1911–12 Professor F. B. Loomis of Amherst College
spent a few weeks in the same region visited by our expedition, but devoting himself to a different subject of study.
Finally, in 1926–27 E. S. Riggs of the Field Museum in
Chicago made a large collection in Santa Cruz and southern
Chubut.

So Darwin, Ameghino, Hatcher, Brown, Loomis, Riggs,
and a good many others, too, have gone rooting around in the
débris of the lost world of Patagonia, and what they have
found is rather startling. If you dig bones in North America,
you find ancient elephants, three-toed horses, rhinoceroses
with paired horns, saber-toothed cats, and the like. They do
not seem to belong in North America, but that is only because

they once lived almost everywhere and happened to become extinct here, and although they are not quite like anything living today, most of them belong to the same families and are not wholly unfamiliar. But down there in South America they found astrapotheres, and homalodontotheres, and sparassodonts, and toxodonts, and pyrotheres, and a lot of other -theres and -odonts with wholly unfamiliar names. It is baffling to try to describe these in terms of any animals living in the world today, for they have left no descendants nor even any distant relatives. They are not like the extinct animals of any other part of the world. To describe them you have to start from the ground up, or to compare them with half a dozen different animals at once, and then add a few original touches, like the fantastic combination beasts in children's stories. An astrapothere, for instance, was about the size of a rhinoceros but didn't look a bit like one. He was four-footed and had an ungainly body as if his inventor had not been able to draw well. He had a snout something like that of a tapir, and tusks something like those of a wild boar but bigger and not so curved. He had no front teeth in his upper jaw, and his lower front teeth were neatly scalloped; if they weren't so large they would make elegant lodge emblems. He probably ate bushes, twigs, leaves, and things like that.

The reason for this startling originality in the ancient life of South America is fairly clear. During most of the Age of Mammals, that is, during the last 60,000,000 years or so, when our animals of today were evolving into what they are, North America, Europe, Asia, and Africa were joined together off and on. When new kinds of animals developed on any one of them, these beasts would sooner or later get the urge to travel and in most cases succeeded in finding a land bridge and eventually in spreading over all of these continents. If the new type of animal was unusually intelligent, or fast, or otherwise well fitted to cope with things, it was successful and became more or less world-wide in its numerous varieties. If it were less well equipped, some enemy or competitor would come wandering along from some other continent and it would lose out and soon become extinct, without having time to become anything extremely unorthodox. So it happened that in spite of numerous local differences, these continents did develop more or less the same general type of animal life.

South America was different. At one time it, too, had a land connection with the rest of the world, but that was a very long time ago, before any of our modern animals had arisen, and the creatures that then wandered into South America were very ancient and primitive. They were stranded there. The crust of the earth, always restless, buckled and sagged a little, and near the beginning of the Age of Mammals South America became an island continent. There was no Isthmus of Panama, nor any other land connection with the other continents. So far as animals that can only travel on solid land are concerned, it was completely isolated. The ones that were there could not get out, and none of the modern types developing in the rest of the world could get in. This condition continued for a long time, for millions of years, and the mammals of South America, undisturbed by outside influences, evolved in their own way and into peculiar types never found on any other continent.

Finally the earth's crust heaved a little more—these changes are usually exceedingly slow, but with millions of years to work in they achieve great results nevertheless—and the Isthmus of Panama arose, a land bridge between North and South America. For a time it was probably even wider than it is now, and in any event it permitted land animals to wander back and forth between the two areas which had been separated for so long. Down from North America went cats, dogs, bears, mastodons, horses, peccaries, camels, and a whole host of creatures that had been developing in the rest of the world while South America was an island. Some of these, like the cats, dogs, and bears, survived in both continents, although generally in somewhat different forms. Others, although originating in North America, later became extinct in the north and survived only in the south. Thus it is that some species which we think of as typical of South America, such as its camels (llamas, guanacos, etc.) are really natives of North America. Others, like the mastodons, became extinct everywhere.

South America sent us some animals too, but not so many. A few, like porcupines or the Texas armadillo, are still with us, but others, like the ground sloths and glyptodonts, have since become extinct.

This irruption of animals from North America spelled the doom of most of South America's native fauna. In their continent-asylum they had never had to face such fierce and efficient competition, and relatively few of them were equal to it. The northern animals had already faced rivals and enemies from Europe, Asia, and Africa and had survived. They had learned to run a little faster, or to think a little better, or otherwise to take care of themselves a little more efficiently than the South American stay-at-homes, and it was not long, geologically speaking, before the latter became extinct.

I am glad that South American politicians do not know about this. They would blame it on Yankee Imperialismo, too.

It was about or shortly after the middle of the Age of Mammals that the peculiarly South American animals reached their highest development. They had then been isolated long enough to have developed their own bizarre characteristics, and the fatal invasion from the north had not yet begun. The extinct animals of that epoch have also been the best known. Their remains happen to be very abundant, very well preserved, and scattered over a large area, especially in the territory of Santa Cruz. And because they are so abundant and well preserved, more attention has been paid to them. Most expeditions have concentrated on them because there they were sure of getting many and good specimens, and expedition leaders, like everyone else, wish to be successful and make a big showing.

Yet that leaves a great deal of curiosity unsatisfied. Those creatures are all very well and the vast labor that has now made them so well known, among scientists at least, was fine and necessary, but it does not complete the picture. What all students would like to know now, and what I particularly am keen to try to learn, is what the ancestors of those animals were like. Where did they come from? Scientists are agreed that they cannot have originated spontaneously in South America. Did they come from North America in the very remote past? or Australia? or Antarctica? or Africa? or all or several of these? And before they were shut up there in South America they must have borne some sort of relationship to the beasts of the rest of the world. Just what were these relationships? And why and how did they develop their strange characteristics anyway?

The way to find out the answers to these questions, and others related to them, is to look for the remains of older animals. If we could find fossil animals that were living in Patagonia while South America was united to the rest of the world, along about the beginning of the Age of Mammals, or soon after it became isolated, they should go far toward solving these problems. They should not yet have lost definite traces of their origin, and they should show the very beginnings of the bizarre modifications undergone by their descendants.

The study of origins is always peculiarly difficult, and I suppose that is a large part of its fascination for me and for many others. The Ameghinos made a good start at this, as they did at most of the problems of the geologic history of their part of the world. Carlos did find the remains of very ancient animals in Patagonia, and Florentino named and described them. They were very scarce, and their remains were very poorly preserved. Most of them were represented only by isolated scraps or single teeth. From such poor material it was not possible in any case to figure out a great deal about the real characteristics and relationships of these oldest of South American mammals. Florentino was obsessed by that idea of his that all the world's animals originated in South America and these scraps were of crucial importance to his theories. In trying to interpret them he went perhaps even farther astray than usual. Students in other parts of the world saw or suspected that Florentino had not interpreted this part of the history correctly, but they were dismayed by the fragmentary nature of the evidence and the difficulty of studying it or of adding to it. Their very hesitancy was a tribute to the Ameghinos. For a long time other workers were content to collect in the richer deposits of later mammals and no really serious and extensive effort was made to add to or to improve on the collections of these earliest forms.

So it fell to our expedition to take the next step in this important and interesting work. The idea was not particularly original. Its necessity, from a scientific point of view, had struck many students, and in fact the first tentative suggestion which resulted in our work came from Professor von Huene, of Tübingen University in Germany, who was unfortunately unable to carry through his plan for coöperation with us. That such an expedition as ours had not been undertaken before was due to the necessity for a combination of unusual circumstances, which happily fell to our lot. In the first place it was necessary to find a patron who was sport enough to back a long shot like this. We were almost in despair until we met Mr. H. S. Scarritt, who immediately supplied the essential backing for our expedition. Then it was necessary for the Museum authorities to permit such work, from the point of view of exhibition inevitably giving less showy, although from a scientific point of view more valuable, results than the same time and effort expended in some richer field. The authorities of the American Museum made no difficulties in this respect. Finally it was essential to obtain the full coöperation of the Argentine scientific and civil authorities. This, as has been seen, was also forthcoming, largely because of the friendly and disinterested spirit of the directors and staffs of the leading Argentine museums.

Of course we did not expect to solve all the problems that I have briefly sketched in this chapter. Perhaps they will never be completely solved. We did hope to make a substantial contribution to their solution, and so far the results have considerably exceeded our expectations. We were successful in making a large and good collection. It proves to contain a number of animals hitherto unknown, and, what is still more important, to contain good and relatively complete specimens of animals hitherto known from mere fragments. The study of this fine material is still in progress and it is too soon to say very much about the ultimate results. Anyway, that is a different story. Here I have only tried to show just why we were in Patagonia. For being in Patagonia without a reason, even such an involved reason as this, would imply a mental condition to which no one would confess.

Or perhaps Manuel was right.

This has been a long digression but it will be easier to see just what we are doing on this cliff at Lake Colhué-Huapí. It is more than a cliff; it is a long chapter in an only partly legible history.

The first day out of Comodoro, we saw that the land rises from the sea in a series of steep and very irregular terraces to the high, barren, windswept Pampa de Castillo. We went down the far side of this into the broad Valle Hermoso, a more sheltered tableland little over half as high. Across the Valle Hermoso the land rises again, much more gently, then drops abruptly to the basin of Lake Colhué-Huapí, part of the great depression known as the Sarmiento Basin, Cuenca de Sarmiento. This scarp south of Colhué-Huapí forms a belt of badlands, about six or seven miles long in its principal part and continued eastward and westward by somewhat similar scarps. This is the great "barranca south of Colhué-Huapí," nameless but famous.

The shore of the lake itself is formed in part by broad white dusty flats flooded at high water, in part by low, steep banks of clay and soil, and in part by higher vertical cliffs of rock brilliantly streaked red and white. Beyond these shores lies a strip of gently rolling land covered by sand and gravel dotted with barren knolls of somber shale. Above this rises the irregular line of the forecliff, vividly banded and spotted in crimson, orange, yellow, and white. It is clearly stratified, in some places horizontally, and in others at steep angles. Beyond this bright and rugged forecliff is the main barranca, rising to over four hundred feet from its immediate base and nearly eight hundred feet above the lake at its culminating point. It is formed by a great series of beds of volcanic ash, white or delicately tinted yellowish or pink except for a few outstanding strata that have weathered to an orange color.

In this ash series, particularly in the lower part, there are many beautiful minerals. Seams of snow-white or pink-streaked gypsum. Clear crystals of the same mineral, gleaming in the sun. Plates and fancifully contorted forms of white or bluish translucent chalcedony. Odd balls of tubes of chalcedony which often prove to be hollow and may be lined with myriads of small, clear, rock crystals or filled with yellowish, spongy, sugar-like quartz, looking very appetizing but harder than glass. "Desert roses," globules of complexly twinned crystals of calcite and barite, looking strangely like petrified flowers, usually white but sometimes pink or even black.

Closer inspection of the barranca makes one long very literally for the wings of a dove. Foot-by-foot investigation, the task of the bone hunter, often calls for a high order of mountaineering skill. There are steep rock slides, shale slopes where the weathered clay slides off the hard subsurface into an abyss below, vertical cliffs with narrow clefts as the only practicable passes. In the softer rocks the water or recurrent rains have worn numerous caves and dark holes which seem bottomless. The only trails are those of guanacos, crossing the slopes at every practicable point in an apparently aimless fashion. In only one place, where the barranca is lowest, have they worn a real highway, deeply rutted, where many tracks from the Valle Hermoso converge and then descend as a guanaco boulevard to the lake.

The barranca, burnt by the sun and swept by the wind, is not so barren as it seems. On closer acquaintance, it is found to be teeming with life. Up there, where there seems to be no smallest tuft of verdure, there are busy ant hills, numerous spiders and beetles, and stranger insects crawl, jump and fly. Birds are innumerable and ever present: eagles perching on the highest crags, and farther down, in nooks and caves, parrots and doves nesting and other birds seeking a restless living. Field mice and other rodents manage somehow to exist here. Guanacos, pasturing only on the gentler slopes above and below, yet spend much time on the cliff, as if for amusement.

This multitude of living creatures wandering over the surface of the barranca is as nothing, either in numbers or in variety, to the stranger and almost inconceivably ancient population that lies buried, awaiting resurrection, in the beds of ash and clay.

Here the sea once roared, and the deposits of its ancient shore now lie far down, near the shore of the present inland lake; in these rocks are the shells of small queerly distorted oysters and the teeth of sharks like none that now swim in any of the oceans. Then, slowly, the continent heaved and this region rose above the sea. For millions of years it was land and on it were spread out and piled up great thicknesses of volcanic ashes, carried by the wind and by streams from the volcanoes of the ancestral cordillera to the westward. Sometimes the earth yawned, and through fissures there welled up molten floods of black lava. Then the crust sank again, and the sea laid down beds of sand and gravel over the thick ash series. In this younger sea lived the enormous Hatcher's oysters, and in it also swam strange toothed whales and now extinct penguins, some of them taller than a man. Again the land rose and the sea withdrew to the east. After some chapters whose record has been destroyed here, but can be read in other parts of Patagonia, erosive forces, wind, rain, rivers, perhaps also glaciers, shaped the present surface, biting deeply into the beds of rock laid down in previous ages and exposing their secrets.

It might seem that animal life could not flourish in a region where so much radical remodeling of the face of the earth was going on, but these things were slow and seldom really violent. The last paragraph condenses over sixty million years of history, and in spite of their great aggregate changes most of those years were calm, and during most of them this region was peculiarly favorable to animal life. There was a tremendously long heyday here when the region was neither so dry nor so cold as it is now. Then vegetation flourished, even great trees—for there are their petrified trunks to prove it—and there also flourished a host of herbivorous animals, eating the vegetation, and another, but smaller, host of carnivorous animals, eating the herbivores.

This is all recorded in the barranca, and especially in its most prominent and most interesting part, that thick volcanic ash series that is intercalated between the two records of submersion below the sea. These ash beds represent so long a time, from forty-five or fifty to twenty-five or thirty million years ago, that evolution and migration brought about many changes in animal life during that span. The animals whose bones are preserved in the lowest, and hence oldest, strata of this series are all entirely different from those found in the highest and youngest levels. On the basis of these changes, the whole series can be divided into four successive geological formations, each with its typical fauna, and each with a definite designation. This designation is double, one name being derived from a locality where the formation is particularly well developed, and another from one of the most common or most typical animals found fossil in the beds. The oldest deposit of this ash series in the barranca is the Casamayor Formation containing the *Notostylops* Fauna. Next in order come the Musters Formation and its *Astraponotus* Fauna, then the Deseado Formation with the *Pyrotherium* Fauna, and finally, at the top of the barranca and just under the remnants of the second encroachment of the sea, the Colhué-Huapí Formation which contains the *Colpodon* Fauna.

There are three sorts of changes in their animal life that distinguish these successive deposits and permit their recognition. Some of the animals in the older beds lived on, but the time is so great that they evolved into new species, and the descendants in the younger strata are unlike their ancestors in the older. Some of the ancient animals became extinct and left no descendants. And some of the younger animals have no ancestors buried in the older beds, but were immigrants whose ancestors had lived in some other part of South America. In these three ways, life was constantly changing, as it still is today.

The animals buried here would look absurdly strange to our modern eyes. Here in the oldest deposits, the Casamayor Formation, which particularly interested our party, is indeed a Lost World. There are just two sorts of animals that would not look wholly unfamiliar today: opossums and armadillos. These, too, have changed, but relatively little so that even at that remote time these creatures were already quite recognizable.

Among the other mammals the carnivores of that time were not wolves or cats or bears or any of the other sorts now so common, but were marsupials, pouched like opossums and with brains of very poor quality. They must have looked a little like our modern types of wild dogs, and some of them were as big as wolves and had great fangs. Even more common flesh-eaters were some of the reptiles, crocodiles and enormous snakes resembling pythons or boa-constrictors.

As is always and necessarily true in any animal society, plant-feeders were much more abundant than flesh-eaters. Most of these herbivores belonged to the distinctive group of South American hoofed mammals, almost confined to that continent, and now all extinct. Of these the most varied and most common subdivision was that which labors under the unhappily long but rather rhythmic name of homalodonto-theres. These creatures had five toes, each ending in a hoof, stocky legs, long heavy tails, and disproportionately large heads. Although they must have fed entirely on leaves, twigs, bark, and the like, most of them had large, sharp fangs, used, doubtless, for fighting among themselves and against their enemies. They varied in size from little things no bigger than lap-dogs up to the largest creatures of their time, about the size of Shetland ponies. Later there were to be much larger animals in Patagonia, but at this most remote epoch of the Age of Mammals these had not yet evolved and their ancestors were still relatively small.

Then there were delicate and swift-running little beasts known as typotheres. Although they were hoofed, they must have looked, and probably also acted, very much like hares. *Notostylops*, a common mammal for which the fauna is named, also had the strange combination of gnawing teeth and hoofed feet. True rodents, akin to the porcupine and to the guinea-pig, were later to be very abundant in Patagonia, but they appear first in the Deseado Formation, above the middle of the barranca, and are quite lacking at the older levels. Where they came from is one of the things that we would particularly like to learn, but our work so far has cast no light on this mystery.

Throughout the whole thickness of the barranca there are astrapotheres, those ungainly, unbelievable brutes which I have already tried unsuccessfully to describe. They first appear, in the oldest strata, no bigger than sheep and with rather short tusks. As you climb up the cliff, up through time, their evolution takes place before your eyes, until at the top they are great beasts as big as elephants and with savage-looking, scimitar-like tusks four feet long.

The racial span of the pyrotheres was shorter, and their whole known history is recorded in the lower three-fourths of this barranca. When its highest strata were being formed the pyrotheres were already extinct. They also begin as rather small animals, unfortunately rare and still very imperfectly known, and end in the Deseado, the *Pyrotherium* Fauna, as animals nearly elephantine in size and also strongly elephantine in appearance. They had trunks, probably a little shorter than those of modern elephants, and they had tusks, two in the skull and two in the lower jaw. Their grinding teeth had two high, sharp, transverse crests. They had so many features

in common with elephants that Ameghino claimed them to be
the actual ancestors of these, and even the American student
Loomis believed that there is some relationship. Now, however, it is widely agreed, and I strongly concur, that the
resemblance is illusive. Pyrotheres and elephants had similar
habits of life, and so came to resemble each other, much as
the whales have come to resemble fish although they are very
far removed in blood relationship, but the pyrotheres had a
very different ancestry from the elephants. Among other
things, this is shown as a result of our own study from which
it appears that the oldest pyrotheres, which should be closest
to any common ancestor, are much less like the ancient
elephants than are the younger forms.

How all these animals came to be buried and preserved
is graphically shown by the same sort of thing going on today
in the Sarmiento Basin, here at the foot of our barranca and
all around the lakes. The basin is slowly being filled by sand,
silt, and eroded volcanic ash from the surrounding higher land.
Rain washes down sediment, temporary streams are choked
with it, and the wind gathers dust and sand from an enormous
area and deposits much of it here in the lakes or along their
margins. These deposits are forming not only in the waters
of Colhué-Huapí and Musters but also and very extensively
on the normally dry land along their shores, and the deposits
are just as truly geological strata, in spite of their recent age,
as are any similar beds that were laid down millions of years
ago.

And the remains of animals are being buried in them, just
as the remains of the vastly more ancient animals which we
seek in the barranca were once buried. Here birds and
mammals are constantly dying or being killed. Sometimes
the bones are eaten, and usually they lie on the surface until
they disintegrate, but a few happen to fall in mud or soft sand
or to be covered by dunes or by flood deposits before they de-
compose. These are preserved and they will, in the course of
time, become fossils.

The future history can be predicted, for it is exactly what

has happened innumerable times before in just such basins during the great length of geologic time. Eventually the basins will be filled with sediment and the lakes will disappear. In time, thousands or perhaps millions of years hence, with the constant, gradual shifting of the earth's crust and the enduring play of the elements, streams will cut valleys back into the deposits that filled the basin, deposits now consolidated into layers of rocks of varying hardness by the pressure of their own weight and the chemical action of the water standing in or slowly seeping through them. These strata will again be exposed along the sides of the valleys and ravines, and some paleontologist of a higher race in that dim future can here collect relics of the twentieth century and study its quaint, extinct life, just as we are studying the life of the year 45,000,000 B.C. The bones now being buried will then have lost all their animal matter, and the chemical-laden ground-water will have deposited minerals in the pore spaces of their inorganic framework. They will be mineralized or fossilized bones, like those we are collecting.

This vivid present example shows, too, why we so seldom find more than one bone of the same animal, and almost never a really complete skeleton. Under such conditions animals are very rarely buried immediately or whole. Carnivores and carrion-eaters pull them to pieces. The bones lie around on the surface for a time and usually become widely scattered before any of them are buried. Most of them disintegrate. Perhaps a flood washed some of them away. Finally a few of the more resistant parts, particularly the teeth, may be covered over and preserved, but now so far apart that it would be impossible to find them all, or, having found them, to establish with certainty that they did come from the same animal.

A fair representation of the life of Patagonia is being preserved here. Guanaco, ostrich, and sheep bones will be common fossils in this geological formation of the future. In the deposits forming in the actual lakes, fish skeletons will probably not be uncommon. Other creatures will be much more rare, but probably almost every species that lives in central Patagonia, including man, will be represented by some fragments. Yet this will give the student of the distant future a pitifully inadequate idea of the life of South America at the present time, or even of Patagonia as a whole. This is another of the difficulties that beset paleontological research. In any one field, the collector will find only the remains of the animals that happened to live in one restricted area and under the particular conditions obtaining there. For instance, this deposit will contain no trace of deer, and yet deer are fairly common in Patagonia, back in the mountains. Still less will it contain any hint of the rich and very different life of other parts of South America.

So the past history of life on the earth has to be pieced together slowly and laboriously from many finds and from decades of work in different regions. The competent student needs constantly to visualize, and to allow for, the difficulties and the inevitable gaps.

Probably the procedure of hunting for fossils can already be inferred from what has been told of their occurrence, "How do you know where to dig?" is always the first question asked when a new acquaintance hears of my profession. The answer, of course, is that we dig where we see something to dig for. There is nothing esoteric about hunting for bones; we have no sixth sense and cannot see into the earth. The first step is to go where we know fossils do occur or where we think they might. Picking a likely place requires a fairly wide knowledge of geology and close study of any previous records of travelers and explorers in the area to be visited, but thereafter it is chiefly a matter of hard work. We hunt for fossils the way you might hunt for a lost collar button, with the difference that we do not know just what we are going to find and so must be on the lookout for everything, and that we have thousands of square miles to hunt in. Sometimes discouraging days, even weeks, may pass without finding anything, and then we may find a rich pocket where a large collection can be made in one quarry.

Having come to a place where fossils do occur, like this

barranca south of Colhué-Huapí, we walk over it, eyes glued to the ground, examining, as nearly as practical, every square foot of the exposed rocks. When the fossils are small but sufficiently abundant or important to justify the effort, we may literally crawl for miles on hands and knees. Many specimens have already been washed out of the rock and lie loose on the surface. Then we must endeavor to find the exact layer in which they were buried, for more of the same specimen may still be embedded there, or there may be other things there, for fossils often tend to flock together and to be much more common in certain limited strata than in others. A large proportion of the things found are worthless, isolated scraps, not worth saving. Complete teeth are usually worth while, jaws always, and complete skulls are as precious as gems. Skeletons, with rare exceptions not including the deposits we studied in Patagonia, are so seldom found as to be quite priceless.

When a bone has been found partly embedded in the rock, it must be exposed carefully to determine its extent and value, and if it is found to be worth collecting, it must be removed without damage. This is more complicated than it sounds, and success requires patience and experience. If a large excavation is necessary, this is made in the ordinary way with pick and shovel, or even in some cases with a horse or tractor-drawn scoop or with dynamite—these latter were never required in our Patagonian work, however, as the largest necessary excavation was only a few feet deep, and the rock is generally easily broken with a pick. For usual prospecting, for small excavations, and for working close to the bone where delicacy is required to avoid injury to it, a special small pick, pointed at one end and adze-shaped at the other, and small enough to be swung easily with one hand, is used. This is the universal implement of the bone-digger and he feels completely lost without it.

Exposing the actual bone is still more delicate, for, like glass, fossil bone is usually hard but always extremely fragile. This fine work is done with small curved awls or, if the rock is very hard, with slender chisels. The resulting debris is brushed away with a whisk-broom or with a small dry paint-brush. If the specimen is uncracked, firm, and small it sometimes requires no further preparation than wrapping in cotton to avoid breakage in packing. Larger and more delicate or cracked specimens require special treatment. No more of the bone is exposed than is necessary to determine the size of the specimen and its general nature, and it is taken out still embedded in a block of the surrounding rock or matrix; freeing it entirely would weaken it and is such a long, slow operation that it should be done in a laboratory and not in the field.

Thin shellac is poured copiously onto the specimen, hardening the surface and helping to hold the whole mass together. The exposed bone is then covered with Japanese rice paper, also shellacked to make it adhere closely and dry to a thin, hard shell. This serves the double purpose of protecting and strengthening the surface and of keeping the bandages, if used, from sticking to the bone itself, from which they could then be removed only with difficulty and possible damage. The bandages are strips of burlap, three or four inches wide, soaked in flour paste or in plaster, and then pressed onto the specimen with the fingers. These bandages are overlapped and crisscrossed, and may be several layers deep depending on the requirements for strength. When dry, these form a hard protective casing which holds the specimen together and prevents injury to it during its travels back to the Museum. In some cases even this is not sufficiently strong, and then wooden splints are set into the bandages.

When a fairly large block must be removed, for instance a complete skull or several different bones of a skeleton which cannot safely be separated in the field, the upper side is first exposed, shellacked, papered, and bandaged. Then a deep trench is dug around the block, the sides are undercut as deeply as seems safe, and perhaps one or more tunnels are dug under it. Bandages are applied to the undercut sides and run through the tunnels as cinches. When all this is dry and firm, the one or more columns left supporting the block are broken, the block is carefully rolled over, the lower side trimmed of its excess matrix, then shellacked and bandaged in its turn.

With the use of this general method, with some little tricks of technique too detailed to specify here, and with some common sense, it is fair to say that there is no specimen so broken or so fragile that it cannot be preserved and taken back to the Museum without injury. One sometimes reads of marvelous discoveries of the bones of extinct monsters or of men which were so delicate and so old that they crumbled at a touch, preventing the writer from producing them as evidence of his tale. Such things do not exist. No bones are so crumbly that they cannot be collected by an experienced bone digger. Even if they are mere mounds of dust, that dust can be preserved and brought back in exactly the shape in which it was found.

Finding and collecting specimens is the main part of a fossil-digger's job, but if he does only that, his work is very inadequate. Each specimen must be numbered, and in a note-book under each number must be entered the exact spot where the specimen was found, a clear designation of the rock stratum from which it came, and other pertinent information such as the date and the name of the collector. When working in regions that are inadequately surveyed, it often also falls to the bone-digger to make at least rough sketch maps of his travels. He should be a fairly accomplished geologist and should study the composition and structure of the rocks in which his specimens occur, measure the thickness of the strata and determine any tilting or faulting that has taken place. Tasks of this sort are almost endless, and the senior member of such a party sometimes finds that most of his time must be devoted to other things than actual collecting. He always finds that his work does not end at sundown, as the notes for each day must be written or placed in permanent form while they are still fresh and exactly in mind, which often takes until far into the night. Everyone in the party is considered as on duty twenty-four hours a day, seven days a week, and holidays, if any, come only on particularly great occasions or when work is impossible for some reason.

Even when collecting is completed, when the fossils are all securely packed in stout boxes, when local legal requirements have all been met, and when the shipment has actually reached the Museum—even then the work is only well begun. The rest takes place in the Museum, and invariably takes much longer than the time spent in the field. Preparators must clean each bone, removing the encasing matrix grain by grain. If soft, the bones must be carefully hardened, and if fragile, they must be reinforced with steel rods inside them or in some inconspicuous place along the outside. Fragments found scattered must be carefully matched, like a jig-saw puzzle, to see whether they will not fit together and make something more complete. Missing parts must be modeled in plaster, so far as this can be done with no possibility of error, by comparison with other similar specimens.

When this work is done, the specimens must be studied. They are carefully compared with all others known and the name of the animal to which they belong is determined. If they are of some creature not known before, a new name must be given to them and a description published. New or otherwise particularly important things must be photographed or drawn. The age of the strata from which the fossils came must be determined. Detailed reports on the results of the whole expedition must be written and published.

Finally, specimens must be selected and placed on exhibition. Iron supports have to be made to hold them, often a long and difficult task. The exhibition needs to be planned carefully, and comprehensible and enlightening labels composed and printed. Specimens not desired for exhibition must also be catalogued and then stored where they can be found readily when needed for study. In any collection there are numerous specimens, sometimes far the greater number, that are not exhibited. A specimen may have very great scientific value, and yet not have the properties of popular appeal, of completeness, of striking character, or of clarity in demonstrating some special point, that are requisite for appropriate exhibition in a large Museum. The study collections of such an institution usually exceed the public exhibition both in bulk and in value.

Fossil hunting is far the most fascinating of all sports. I speak for myself, although I do not see how any true sportsman could fail to agree with me if he had tried bone digging. It has some danger, enough to give it zest and probably about as much as in the average modern engineered big-game hunt, and the danger is wholly to the hunter. It has uncertainty and excitement and all the thrills of gambling with none of its vicious features. The hunter never knows what his bag may be, perhaps nothing, perhaps a creature never before seen by human eyes. Over the next hill may lie a great discovery! It requires knowledge, skill, and some degree of hardihood. And its results are so much more important, more worth while, and more enduring than those of any other sport! The fossil hunter does not kill; he resurrects. And the result of his sport is to add to the sum of human pleasure and to the treasures of human knowledge.

CHAPTER V

COLHUÉ-HUAPÍ

COLEY had arrived in Comodoro on October twentieth, and when Justino and I went to town two days later we found him there, still convalescent but anxious to get out to camp. The following day was devoted largely to putting the car in better running order. Whether because of the change of climate, the sea voyage, or some other disturbing factor, its performance so far had been uniformly bad, but it later settled down to work and responded fairly well to the unusually severe demands made on it.

Our evening in town was spent at the movies. When we arrived there, we found that a very poor film was scheduled and would have left in disgust had not our host, manager of a local bank, sought out the manager of the theater and complained. The scheduled film was stopped in the middle, and instead there was shown a very old but superior film, a print of which had been knocking around town for a couple of years. What the other customers thought did not appear, for they accepted the change very meekly and without explanation. At the end of the feature, the proprietor came to us and told us in awe-struck tones, rubbing his palms together as he spoke, that to honor us he had a very special treat for the norteamericanos. The theater darkened again and after a short wait to pique our curiosity, there appeared on the screen a picture of Hoover announcing his candidacy for president—an interesting event that occurred some three years ago. This was the most recent newsreel that had reached Comodoro.

This unusual deference from the management may have been due, as we afterwards learned, to the fact that the theater

84

was in financial difficulties from which only our host's bank could extricate it.

The day was darkened for me by belated notice of the death, in California, of W. D. Matthew, a treasured friend and honored colleague, one of the finest men I ever knew and one of the greatest scientists of his generation.

My journal continues:

Camp No. 1
Lake Colhué-Huapí
Oct. 24.

It was raining hard when we got up this morning and prospects of making the trip back to camp seemed rather dubious, but I finally decided to try it anyway. After making last-minute purchases, we got away at 9:30. The rain had not made the roads much worse than they already were and we only got stuck the usual number of times. By our route, it is some thirty leagues (about ninety miles) from Comodoro to camp, which doesn't sound like much but is a very hard day's drive. We had lunch in a little boliche (a sort of inn and general store) on top of the pampa and on the railway, and also stopped in Cañadón Pedro to fill our water tank, the camp water not being fit for drinking. It snowed on the pampa last night, but not at camp (which is considerably lower). Camp looks surprisingly good to me, and I am very happy to be here. It seems like home now.

Oct. 26.

After straightening out records and so on, Justino and I worked at the barranca yesterday. Coley is still very seedy· from his illness and stayed in camp, but today he went out with us, although the climb exhausted him and he was not able to do much. The work is coming along famously and we already have a good collection from the later beds, near the top of the cliff. At noon we were much dashed when we had water all hot for maté and then found that we had forgotten to bring yerba, our staff of life. We came down at sunset.

There is one distant spot on the barranca that always receives the last rays of the sun and which we see as a spot of brilliant orange in a dark landscape when we are on our way back to camp. I hope that there is a particularly fine specimen waiting there for us.

As we drove the car along the dusty desert trail, out of sight of any water, we came across a large live fish flopping about in the middle of the road. Bewildered but grateful, we picked it up for supper and drove on, only to encounter another a few yards farther along, and then another and another. We finally had about a dozen, and I have never seen such good fishing as in that dusty trail. Perhaps I should stop the story here and add another and (as far as it goes) perfectly true item to esoteric lore. Isn't it Charles Fort who makes so much of the mystery of life as exemplified in such apparently inexplicable occurrences? Unfortunately for mysticism, Don Francisco has a fishing camp on the lake and just before we came by he had driven along this track to Casa Ramos with a load of fish. The bumps did the rest.

The lake is teeming with fish and they can be caught by the hundreds with nets. This would be a good source of income for somebody if there were a market. Unfortunately there is only Comodoro and it requires an unusual combination of cold weather and fairly passable roads to reach there before the fish spoil. Furthermore, the people here do not really care much for fish and eat very little of it even when it is available. There are three kinds of fish in the lake: pejerrey de agua dulce, similar to the marine pejerrey which is also excellent eating, trucha, which is not a trout despite its name, and bagre, which somewhat resembles a catfish but is, to my taste, better eating. From a gastronomic standpoint (the only one I take) the pejerrey is the best of the lot but all are good.

[I here interrupt the journal for a moment to confess great shame as a supposed scientist. Months later I told an ichthyologist at La Plata about these three sorts of fish and he became much excited and berated me very soundly and deservedly. It seems that this pejerrey, of which we ate so many specimens and did not collect one, is undoubtedly an unnamed species, new to science.]

Oct. 27.

As I write this, just before going to bed, the Patagonian wind is living up to its reputation. Waves roar on the beach. The tent billows and snaps and then strains with the guy ropes humming, but perhaps it will stand up. The strongest wind I have ever seen was blowing on the barranca today. To climb over the crest I had to crawl on my belly and in a less cautious moment I was knocked down and almost blown over the cliff. At one time going into the wind down a slope too steep to stand on at all ordinarily, we could walk leaning forward at an apparently fatal angle, supported by the constant gale in our faces. There could hardly be a more curious sight or a stranger sensation. Just there the wind was blowing such large pebbles that we had to remove our goggles for fear of their being hit and broken.

My present field garb, ridiculous to others but practical for me, consists of boots, khaki riding breeches, flannel shirt, canvas hunting coat, air-tight goggles, and a beret or boina. In this it is possible to work in spite of the wind. Work is not simple, however, at times like today when the wind blows away even tools, and the fossils we are trying to collect.

Coley put in a few good licks, but is still very weak and had to quit early. What with having to leave his bride, his severe illness, and other things he is not quite as favorably impressed with Patagonia as I am.

We put in a good day of work. We found traces of the last fossil collectors—Riggs' party some six or seven years ago —and even came across a specimen which they had started to collect and then had left for some reason. They worked altogether on the richer upper beds, whereas we are about done with them and are now going to concentrate on the older and rarer things near the foot of the barranca.

European hares were introduced into Patagonia a few years ago by some very misguided soul and now have overrun the country, threatening to do in the native and more valuable fauna.—This apropos of the fact that we had one for dinner tonight and that it tasted very good. Odd how the mind keeps reverting to food out here!

No one sat around after eating, all having had quite enough for one day. The wind is very trying, but we manage to keep on good terms and in good humor.

Oct. 28.

The wind was even more violent today and my eyes were swollen nearly shut from the sharp grains of volcanic ash blown into them yesterday, so we stayed in camp. There was plenty to do, as several of our specimens were not completely prepared for packing and my notes and geologic sections, which consume a great deal of time, needed copying and revising.

Justino went off and bought us half a sheep, also picking up a live armadillo, pichi, on the way. It is a gravid female, intensely upset at her captivity and sulking in a pail of dirt at this moment. A shepherd came by on horseback and told us there were some bones a few kilometers away. We went there with him, but the "bones" were bits of petrified wood of no value.

Oct. 31.

The last two days and today were spent working at the barranca. We have now transferred our attentions to the old beds in the lower half of the cliff. Here there is a thick bed of pure gray volcanic ash which is fairly hard and very tough when dry but which washes out easily when wet. In this way the occasional torrential downpours have honeycombed the ash bed with small caves and channels. In one of these caves, in a block fallen from the ceiling, Justino and I found the skull of a fossil crocodile.

[This specimen later turned out to be very important. It is of a genus and species new to science, and it also indicates the ancestry of the caimans and jacares now living in South

America and suggests that they probably wandered there from North America some sixty million years or more ago.]

We had lunch in the crocodile cave today. Justino, who prospects around like an overgrown fox terrier, but with very good results, has wormed his way into all the little caves and holes around here. He found one excellent specimen in a tiny crack in the ceiling of a cave so small and narrow that it had to be enlarged with a pick to be passable at all. He is an amazingly good prospector and seems to smell out specimens in the most unlikely places.

The armadillo has been staked out with a cord tied to one leg, and has dug herself a hole and retired permanently to it. A most unsocial creature.

Nov. 1.

Justino, Manuel, and I went off, leaving Coley to hold down the camp and rest himself. The lads dropped me off at a strip of badlands on the slope of the Pampa Castillo, a few miles from the Valle Hermoso railway station and they went on to 112, on the Pampa, to make a few purchases and mail letters. Hence they came back by way of Cañadón Pedro, where they stocked up on drinking water and got a lamb. I found a few specimens but didn't accomplish as much as I wished because my eyes swelled shut after a couple of hours in the sun.

Guanaco are fairly common in that area—there was one troupe of thirteen. On the cliff there were two of the very handsome big black and white eagles, aguiluchos. They were disturbed by my climbing their perch and the male amused himself and got on my nerves by going to a great height and then dropping straight for me, turning at the last moment and buzzing by a foot above my head with the speed of a cannon ball, then banking and soaring back to repeat the maneuver. He did not strike, but whenever he swooped I could not help wondering just what his intentions toward me were.

I was also entertained by another bird which Justino calls

a chuchumento—I cannot guarantee the name as Justino hates to admit that there is anything about this country that he does not know and so occasionally invents answers. This chuchumento, or what have you, builds a large nest of firmly intertwined twigs in thorn bushes several feet above the ground. The nest itself is spherical and is completely enclosed except for a small entrance tunnel about two feet long. This is set with sharp, fresh thorns so as to be almost impregnable.

Incidentally I was much surprised the other day to find the fresh, recent bones of a very small opossum. I had no idea they ranged this far south, and in fact they must be very rare and probably nocturnal, for Justino had never seen or heard of such an animal and insisted that the bones must belong to a weasel.

Nov. 5.

Coley is well now and has been working full time and making some fine discoveries the last few days, and of course Justino is in good form. I have been out and have done what I could, but that was not always a great deal as my eyes are still very bad.

Today Justino and I went down to the far end of the barranca and walked back on the cliff, a good trick involving much climbing and walking on narrow ledges, jumping chasms, sliding down clay slopes, and in general a combination of Alpine and field sports. The work here is pretty well laid out now, and I think it will take till the end of the month to clean up the things within reach of our present base camp.

We had cold pichi—armadillo—meat for lunch. The meat is dark and rather pleasant, although it is very greasy. I would not care for it as a steady diet. The ostrich laying season is in full swing and we gathered some eggs. It is possible to select the fresh eggs in a nest because when new-laid they are green and have a limy crust and as they get older the color fades and the lime is worn off. There were

forty-eight eggs in one nest. To return to my usual gastronomic theme, these eggs are extremely good food. They
make excellent omelettes, tortillas, and are even better, and
more exotic, roasted in the native manner. The top is carefully cut off and part of the white is removed and replaced
with sugar. The contents, still in the shell, are then mixed
thoroughly and the whole thing banked in hot ashes until
cooked through. As I write it, this does not sound very
enticing, but in fact it is one of the best things I ever ate.

In one place the eggs were neatly piled up several feet
below the nest which, as often, was on a little knoll. Explaining this stopped Justino for only a moment, then his inventive
genius came to his aid:

"You see, doctor, one egg rolled out and the bird was too
lazy to put it back in the nest. So he moved all the other eggs
down to it."

"And how did he carry the eggs?"

"Oh, he tucks them under his chin and holds them there
with a loop in his neck!"

If Justino could only understand English I would tell
him about the Side-Hill Squeegee of my native Colorado
mountains.

Nov. 6.

After having breakfasted on armadillo sauté, we went out
to the barranca and found his oldest known ancestor, proving
that armadillos were in existence, and pretty much the same as
they are today, some forty-five million years ago. To all his
other strange characteristics the armadillo adds that of being
a "living fossil." He has hardly changed at all since the
Eocene Epoch and should by all rights be an extinct and prehistoric critter, but in his quiet and dumb—how dumb!—way
he keeps on anachronistically surviving. Apparently in the
struggle for existence a thick skin serves just as well as
brains, or better.

Speaking of food, we are almost out of meat and cannot
get a sheep till tomorrow. Coley shot an abutarda near

Francisco's fishing hut, but it was on the other side of a small arm of the lake. He started after it in Francisco's boat, but the boat sank when halfway across. This boat is one of the world's sad sights, made out of tin cans and a couple of old planks, with cracks as wide as a finger. Francisco goes out in it on this very tempestuous lake, but he can and does swim. When Coley finally got across the bird had recovered and flown away. We had ostrich egg omelette for supper.

This very peculiar desert even presents difficulties in navigation!

Nov. 9.

The work has been going along smoothly, with the usual nasty weather and good fossils.

This morning just before dawn, when the wind had been almost quiescent for a change, the gale suddenly hit my tent with a resounding whack and then rain pelted down, the drops drumming on the taut canvas like machine gun bullets. I finally dozed off and only a second later there was a hand-clap outside and I heard Justino's cheery (but also sleepy) voice calling "Doctor! Linda hora pa' levantarse!" I found his belief that the hour was pretty for getting up difficult to share, but crawled out.

Then there was a clap of thunder and it began to hail, large stones carried along at tremendous velocity by the wind, now of hurricane force, so that it was positively unsafe to venture out of doors (or tent-flaps). These hail storms, which have been frequent, are quite annoying, but seldom last very long. This one let up soon and we did our usual stint at the barranca.

Lately I have been noticing some very strange birds from a distance, and today, getting a closer look, was amazed to see that they are parrots. What on earth is such a tropical-looking bird doing in this raw, cold desert? They are about the size of crows and they build nests of twigs high up on the cliffs—they are called cliff parrots, *loros barranqueros.* All day long they fly about us, very characteristic with their long tails, short necks, and large heads. This odd stream-lined body form makes them singularly like flying fish as they pass overhead. They are greenish but not brightly colored. Their cry is a very ugly coarse raucous squawk. Justino says that they can be tamed but do not talk.

As we came home the sun went out suddenly and the whole land turned a sinister gray, dark and light but without a spot of color. Streaked vicious clouds poured over us like a flood from the west. The moon rose yellow through the last band of clear sky. Rain began to patter, then to pour. Surf is roaring again on the shores of the lake. Beyond this element-tormented spot lies vast, desolate Patagonia. Beyond Patagonia lies the world of seas and plains and mountains, for complacent thousands of miles. Beyond the world wheels the dusty universe.

So, chastened, to bed.

Nov. 10.

Today we had a little taste of civilization, if the word can be stretched to include Colonia Sarmiento. In spite of the all too frequent rain and hail, the lakes have gone down and travel is incomparably easier; we made Sarmiento in only two hours. After lunching at the now familiar Rojo Hotel and transacting our business, I got a haircut, a badly needed luxury, but distressed the barber by making him leave my beard. Once past the painful early stages a beard in camp serves both as an amusement and as a comfort.

Everyone in Sarmiento was delighted to have someone to talk to, and in all the stores and at the inn I had long conversations, but dull ones. Justino, much at home, disappeared and lunched with one of his numerous brothers or cousins.

Wool trains are assembling in Sarmiento, preparatory to the hauling season. Some wool is now taken to the railway or the coast in motor trucks, but the greater part of it, especially from back in the cordillera, still has to be brought much or all of the way in wagons, or *chatas*. These are very stout and very high, often with enormous hind wheels, up

to ten feet in diameter, which enable them to negotiate the
deep sloughs and rutted tracks. Wagons, horses, mules,
harness, and all manner of paraphernalia are piled up all over
town, with the carters putting things in order, gossiping,
sleeping under the wagons, cooking over open fires. These
carters are a rough lot, Indians and mongrels of all sorts,
many of them down from parts of the cordillera that are still
marked "unexplored" on the maps.

We need hay to pack some of our fossils, and Sarmiento,
with its marshy silt, is the only place where any is to be had,
so we stopped at a farmhouse on the way out and bought a
sack of it. It was sold to us by a funny gnome-like little
man who dragged out a few phrases of quaint, rusty English
from a dim recess of his memory. He is a Welshman who
wandered down here from the Chubut Valley when he was
young. He has forgotten practically all of his English and
most of his native language, and even seems to recall his
own origins with great difficulty. Indeed, his mind seems
to be slipping out of contact with the world about him, leaving him with a vague, perpetual smile as if he were living an
inner life based on a private joke.

So back to camp, and tomorrow to go on with our work on
the barranca.

Nov. 15.

The servant problem has entered Patagonia. Manuel has
been getting more and more sullen, and he never was more
than barely passable as a cook, so it finally seemed that he
would have to go. This morning Don Francisco came lurching
up in the ancestral Ramos family flivver to take him away.
We parted with mutual and entirely insincere expressions of
high esteem.

The survivors breakfasted less greasily than usual, with
Justino doing the honors at the stove, and then Justino and
I went off to Sarmiento to get a new cook. Cooks are hard to
find, especially at the busy season of the year, but we located
one Ricardo Baliña who is well recommended and who seems

to be a much more cheerful and willing sort than Manuel. We are slowly working away from Portugal: Baliña was born just across the border in Spain. He came to the Argentine when he was ten and is thoroughly criollo now.

When we first saw him, Baliña had just come in from a particularly inaccessible spot called Laguna de los Palacios, where he lives (rumor hath it) with a slattern Indian woman. His hair hung down in long curls and he had a fuzzy margin of whiskers, but he celebrated by shooting the works in a peluquería and came out clean-shaven, his hair short and plastered down, and smelling strongly of synthetic violets.

I had some difficulty in getting Baliña's last name straight as he is either illiterate or barely literate. Even in correct Spanish there is little difference between *b* and *v*, and the more ignorant use them almost interchangeably or, if they do differentiate between the two, often do so incorrectly. If they must distinguish, as in spelling a name, they call *b* "big *v*" (or "big *b*") and *v* "little *v*" (or "little *b*"). Spanish is supposed to be written as it is pronounced and hence to spell itself automatically, but this does not work out in practice and the common people have about as much trouble with spelling as do ignorant English-speaking people. They not only mix up *b* and *v*, but also *ll* and *y*, which are pronounced exactly the same in the Argentine and most other parts of Latin America, as are *s*, *z*, and soft *c*. The silent initial *h* also introduces difficulties. Then, too, these people do not pronounce common words fully, often saying "pa" for "para," for example, and are inclined to spell as they speak. Justino reads well and speaks grammatically, but his written words are incomprehensible except by reading them aloud and listening to the sound.

Coming back across the flats there were, as usual, numerous birds: ducks, abutardas, herons, and ostriches. Some eggs are already hatching, for we came across an old male ostrich with a mob of chicks. We tried to catch one, but they are several days old and were too quick for us. We do have one ostrich chick that Justino caught a couple of days ago, and which we are keeping in camp and have christened "Charita," the local word for ostrich chick. Somewhat older birds are *charas* and those full grown are *avestruces* or *choiques* (an Indian name). Our Charita is already quite domesticated and seems to be happy and thriving. He is now contentedly sleeping under an old sack in the cook tent and I shall follow his excellent example—under a quillango in my own tent.

A moment ago the stars were beautiful, Venus setting in the west, the Southern Cross a few degrees up and an oddly inverted Orion rising in the east. It is still strange to see familiar constellations, and also the moon, apparently upside down as we here look northward toward them instead of southward as at home.

Charita is dead! He passed away without pain in his sleep last night. This morning when the sack that was his bed was turned over, there he lay curled up naturally, but no happy whistle greeted us and he (or can it have been she?) was stiff and cold. The diagnosis is kerosene poisoning. Decidedly catholic in his tastes at all times, he yesterday found an unprotected can of kerosene while the motor was being cleaned and he drank deeply of this exciting new beverage. He pulled through the day, with only an occasional pause and an expression of bewilderment, apparently wondering what was going on inside him, but during the night it got him.

Florrie, most recent of the several armadillos that we have had, has also departed, not this mortal sphere, but to the wild again. With our tender hearts, we gave her an old sack to sleep under and the ungrateful wench piled it in a corner of her box, climbed over it, and hit for the open spaces. Perhaps she was gifted with prescience, for her sullen disposition had doomed her to the oven. Her only reaction to our most friendly advances was suddenly to uncurl, emitting at the same time an explosive wheeze, a very startling maneuver like having a mild cigar explode in one's face.

After the prescribed period of mourning for a dead charita

and a thankless pichi, we went off to the usual day's work at the barranca.

We were just finishing lunch when a *chulengueador* (hunter of young guanacos) came riding up on a horse completely white except for its eyes, followed by a lean dog completely black. His equipment was typical. The native saddle consists chiefly of two leather-covered cylinders, placed longitudinally one on either side of the horse's spine, on top of which are strapped several sheep skins, with a carpincho hide over all. The carpincho is a large native rodent (in fact, the largest of all living rodents) of northern Argentina. The tanned hide is very soft and porous and is so cool and comfortable for riding that even the poorest horseman usually manages to buy one, although they are expensive. With this type of saddle, stirrups are hardly necessary and generally there is only one, used for mounting but not while riding, although some horsemen have two. The bridle is homemade, painstakingly plaited from leather and often with four lines, all held in the hands or two (from the halter) laid over the horse's neck and the other two (from the bit) held in the hands.

The chulengueador, like almost all horsemen in this region, whatever their business, carried no firearms but had the usual large unsheathed knife stuck vertically into his wide sash in the middle of his back and the equally omnipresent boleadora wound around his waist.

He stopped and passed the time of day with us, of course, taking out his knife and falling to on the little that remained of our lunch. As he ate, he talked, and when the conventions had been satisfied he told us a story.

"In the sierra," he said, "many years ago, men noticed a strange phenomenon. On black nights when the stars were obscured and the wind howled and rain fell on the bleak heights, far away a ball of unearthly light was seen to rise high in the air.

"It moved fitfully through space toward another mountain side, then glided down into the valley, where it hovered for a moment. Then it disappeared into the earth. This happened many times, and men were afraid, for they knew that there were strange things and evil things in the mountains at night.

"Finally they could stand it no longer, and they knew that they must learn the truth whatever might befall them. Painfully they climbed where there were no trails and where it seemed that man had never been before. They came to the place where the ball of light each time started its journey, and there they found a tall tree, the largest in the whole sierra. About and between its roots were great rocks of a kind foreign to the locality and heavier than two men could lift, although the tree had twined its roots about them and pried them apart.

"With immense labor the men tunneled in under the tree and under the rocks, and below them they found the body of an Indian chief. The tree had rooted and grown from the middle of his belly and had flourished on the fat of his intestines. On tempestuous nights, the tormented soul of the Indian chief issued through the tree and clothed in a ball of light it went to the graves of two brother chiefs, whose mouldering bones were duly found where the ball of light was wont to disappear.

"The night after the grave was opened there was a terrific storm and the great tree was overthrown. The ball of light was never seen again, but of the men who found its source, each, in a separate place and at a different time, came to a violent and painful end."

My turn to entertain having come, I told the chulengueador that the spot where we were sitting had once been under the sea, that even before that it had been covered with hot ashes spewed out by distant volcanoes, and that I had today been digging up from among those ashes the bones of an armadillo forty-five million years old.

The chulengueador looked hurt that I should take him for a child who would believe such a tale, muttered an oath under his breath, got up, and forthwith left the gringo liar without any farewell.

We went back and worked until sunset, then home in the crimson light. I cannot go on gushing about each sunset like an old maid on a determined tour, and trying to describe one in words would be like trying to set the taste of old wine to music. But I am sure that nowhere on earth can there be such a gorgeous, brilliant, and varied display of light, color, and cloud forms as we have each evening here in Patagonia. Each one is beautiful, and each is different from any that went before or will come after, some flaming and garish, others mellow and flowing, and all including every conceivable tone and atmospheric feeling. Tonight as we came home it looked as if Pico Oneto, to the east, were looming behind a translucent veil of luminous crimson powder.

Once or twice I have remarked to Justino that the sunset was rather decent that evening, and he has invariably replied, "Yes, it's all right," without bothering to turn his head. He has seen them too often to notice them.

Nov. 19.

Aside from the usual barranca work, adding steadily to our scientific results but not otherwise particularly interesting now, Don Francisco furnished the only diversion today. His Ford, which is so decrepit that it only goes by the grace of God anyway, has been running without oil for some time and he finally decided that he should do something about it. So he drove to camp this evening to borrow some oil (which, like other minor borrowings, he will repay with fish) and also to invite us to an asado tomorrow. We accepted eagerly.

Having more time to write and nothing more to record of daily happenings, I will here insert a short treatise on the boleadora, which is the implement commonly used for hunting:

The boleadora was the one great invention of the pre-Columbian Indians of the Argentine. A few boleadora stones may be found at almost any old Indian campsite, and they are common Indian relics throughout the country. We have already found several. The Europeans accepted this invention with great willingness and boleadoras are still in daily and almost universal use.

A boleadora consists of two or three weights or bolas strung together by leather thongs, which are usually plaited from the neck skin of adult guanacos. The Indian bolas were made of stone, with the thongs tied in a groove around them, and these are still occasionally used. Now, however, the bola usually consists of a sewed, spherical leather bag with the thong attached at one end. The bag may contain a stone, or pebbles, or shot, or even mud. I have been told of an ostentatious gaucho who had fine ivory bolas, and of an Indian chief who struck it rich somehow and could think of no better way to spend his money than by buying hollow bolas of pure gold (which were, however, stolen from him the week after he got them).

Boleadoras vary a great deal but in this region three main types are recognized: the chulenguera, the avestrucera, and the potrera. As their names imply, these are primarily intended to catch young guanacos (chulengos), ostriches, and horses, respectively, but they are by no means confined to those purposes. In Comodoro I was told that one rancher had recourse to boleadoras to catch his children when a priest unexpectedly arrived at the estancia and he wished to have the children baptized. It is only fair to add, however, that the rancher himself denounced this story as a malicious and willful libel—he says that he did not use boleadoras but a lasso.

Each type of boleadora has one small bola, the maneja, which is held in the hand. The chulenguera and potrera have two larger bolas, in addition to the maneja. Each bola has a thong the length of which is about half the extreme span of the owner's arms, and the ends of these three thongs are tied together. Both types vary in weight following individual preference, but the chulenguera is lighter than the potrera. For very large animals over-sized potreras are occasionally made with the three bolas together weighing up to five pounds. The avestrucera has only the maneja and one larger bola and the thong uniting them may be longer than the span of the user's arms.

In practice, a boleadora is whirled above the user's head and then launched at an animal's legs or neck. As it goes through the air, the bolas separate and whirl like chain shot, and if any part of the thongs touches the animal the rest of the boleadora winds around it. If around the legs, the boleadora effectually ties the animal up, and some animals will not run with boleadoras around their necks, even though apparently quite able to do so. The Indians hunted on foot with boleadoras, but now in hunting they are almost always thrown from horseback. Ordinarily the boleadoras themselves do not injure the animals at all—they are commonly used for catching valuable horses, as are lassos in our West (and also in parts of South America). For hunting, however, they do bring the animal to a stop so that the hunter can then walk up and cut the animal's throat. This is the common procedure in most of the hunting in Patagonia, and hunters seldom use firearms. Shotguns are used for flying birds, which are too difficult to catch with boleadoras, and most Patagonians have rifles; but, so far as we observed, these are more for display or for the pride of possession than for use, as they seldom have any cartridges for them and much prefer to use boleadoras when possible.

Nov. 20.

A nasty day of weather, wind, rain and hail, but that isn't news. After some work around camp, we all went off to the asado. An asado (the word simply means "roast") is a grand social function here, almost the only one in the lives of the country people. Neighbors gather from scores of miles around. They gossip, drink, eat, and more than incidentally do a great deal of work for their host. The idea is that of the old corn-husking bees, giving opportunity for a party and social contacts and also getting work done more quickly and pleasantly. In this case the work was the marking of this year's lambs on this range. The tails are clipped, the ears slit with the owner's mark, and the males are castrated. For ease in distinguishing them at a distance, the females' tails are cut very short, the castrated males have about half the tail cut off, and the whole males are left with whole tails.

When we arrived there was a milling mob of sheep in the several brush corrals. Dogs were barking and nipping legs to keep the sheep in order. The men were sorting them out, driving them from one pen to another and selecting and operating on the luckless lambs which were bleating and baaing madly. Justino and Baliña plunged in and helped, but Coley and I felt that we could help best by staying out of the way.

Meanwhile in the windowless, one-roomed mud house the asado itself was being prepared. This consisted of a whole lamb, split down the underside, spread out flat, and fastened to a spit with a cross bar. The fireplace was simply that part of the room that happened to have a piece of tin in the roof instead of rushes. The fire was built on the floor and the spit stuck into the ground so that it leaned over the coals. From time to time the roast was turned and basted with *salmuera*, a concoction for which each cook has his own secret recipe but which ordinarily consists of water, vinegar, salt, black pepper, chile pepper, onion, garlic, and anything else that happens to be handy. Being simply poured over the meat while it is cooking, it imparts a flavor actually quite delicate in spite of the potency of the ingredients. Incidentally this is the way we often cook our meat and it is delicious.

When the roast was done, the men came in, to the number of about twelve. All our neighbors, and some others—Don Francisco, Córdoba (from whom we have been buying our sheep), the chulengueador who visited us the other day at the barranca, the man-with-a-horse (far from distinctive designation for a man whose name I have never heard but who drops in on us now and then), the boy-with-a-horse who sometimes accompanies the man-with-a-horse, Castaño, and two other men and another boy whom I never saw before. Good form does not necessarily require introductions out here, and it is quite possible to know a man fairly well without knowing his name.

Castaño is the most picturesque of this lot. He looks as if he had been made out of the spare parts of several people of different dimensions and races. His eyes are sharply crossed, his complexion is deep mahogany, and his disposition is very cheery. Whether Castaño (chestnut) is actually a name or whether it is merely applied to him in description of his color I have not learned.

Our host had put on a pajama jacket for the occasion, which is considered very chic for outdoor wear if not a trifle too high hat, but the others had just come in their business clothes. We did feel a little badly dressed, as we were the only ones who had not worn socks, and had to console ourselves that it was better no socks at all than socks like some of those that made their presence felt there.

The roast lamb, still on the spit, was stuck in the middle of the circle of guests. Each man drew his knife and dashed for the roast. The first man got the kidneys, then the ribs disappeared, then the legs, joint by joint. When we were through the large lamb had been reduced to a small pile of clean bones and, although I should not like our host to know it, I am sure that we should all have relished a good deal more to eat. We also had a loaf of bread about two feet in diameter and a large washtubful of lettuce—both unusual luxuries, the bread provided by Córdoba, who has a wife (but of course did not bring her to this social function), and the lettuce by Don Francisco, who has a tiny garden patch on the shore of the lake. A good time was had by all, and we went home finally, tired but happy, from our first party.

Now I am sitting in our cook tent with Baliña babbling beside me. So far he is very satisfactory, a good camp cook and good company. Manuel used to get angry and sullen when Coley and I talked English in front of him, as we frequently did. The present incumbent is clever enough not to be angry but to get his own back by talking very colloquial Spanish as rapidly as he can whenever we baffle him by making a remark in English, and by inventing weird and unusual expressions to puzzle us. Butter is "horse-grease," flour "sand," salt "sugar," water "brandy," smoking "throwing ashes out of the volcanic passages," and so on. He thinks it is immensely funny that we are looking for bones here, and every morning he asks us to bring him back a live plesiosaur. All of which is pretty amusing, but will doubtless call for some suppression when his merry quips begin to lose their freshness for us.

Nov. 22.

This day is memorable because we finished work at the barranca and because we had a wild chase after an ostrich, the moral of which was that ostrich hunting cannot yet be successfully or at least efficiently mechanized and that horses and boleadoras are still best.

By noon we had finished the barranca, proper, on which we have worked for so long. At last practically every inch of it has been examined, much of it on hands and knees, and all the visible fossils of any value have been collected. After lunch we went over to a smaller patch of badlands south of the Casa Ramos, and we had finished this and started home when we startled a solitary ostrich. Several distant pot shots with the rifle disturbed him only slightly, but as we approached he started off at a lope and we followed in the car. Another shot hit him (in the leg, as learned later) and dropped him for a moment, but he was up almost immediately and off again, definitely alarmed this time.

The race that followed was epic and would have been prime entertainment for anyone but the ostrich or us. Bird versus motor car, five miles, cross country (and how cross!). The jolts are frightful here when the car is barely crawling, and we went almost full out, the wheels touching the ground seldom but hard. I was hanging on the side, and the truck several times slapped me hard on the ear, which is still swollen. Away went avestruz like a race horse, and after waddled, leapt and banged the car like nothing on earth. After a couple of miles the unhappy bird began to tire from his wound, and would stop to rest whenever he had gained a few hundred yards, a circumstance which alone enabled us to keep up. Finally we came upon him resting in a little water-course. Justino gave him both barrels with the shotgun, but he was up and away at once and disappeared over a hill. Then we gave him up for lost, as the car was too overheated to continue the pursuit at once, but when we could go on we found the poor avestruz only about two hundred yards away, stone dead and full of lead.

Justino cleaned him on the spot and we carried the carcass in triumph back to camp where this same versatile Justino cooked him in Tehuelche Indian style. Neck, wings, and legs are removed and the body is boned without cutting the skin except along the belly. The meat side is then covered with spices and tied up into an air-tight bag, with the skin side out and the inside filled with hot stones. Steam distends the skin and cooks the flesh. Finally the whole thing is carefully placed in live coals to crisp the outside.

This delicacy is exotic. Having eaten avestruz à la Tehuelche is doubtless one of those things that place a man apart as one who has lived extensively—and suffered. Like the bear barbecue of the Rocky Mountain or, still more, the fried rattlesnake of the western plains, this dish is an experience not to be avoided or forgotten, but it is not food. The flavor is high and not quite pleasant, and the consistency is that of somewhat languid shoe leather. At least, so it is for me. Justino relishes it, and Coley seems quite taken with it also.

Nov. 23.

This day falls into observations under several headings:
Geological: We worked some miles distant, but still within reach of our base camp by car, in a cut where the railroad from Comodoro to Sarmiento descends from the high and dry Valle Hermoso into the Cuenca de Sarmiento. Here our fossil-bearing strata are well exposed again, and we obtained several good specimens.

Zoölogical: As we lunched, a band of seven or eight guanacos nearby began suddenly to run madly around in circles, throwing up great clouds of dust and whinnying peculiarly. After a dozen or more rounds, they went off at top speed and ran up and down the precipitous barranca several times before they finally disappeared. Justino says they do this whenever a female drops a chulengo. This may be so—"as for me, I prefer not to affirm."

Meteorological: Wind, or better, WIND.

Social: We had no sooner returned to camp and started on our evening maté than a car drove up and Dr. Feruglio got out. He is a geologist, from northern Italy and very anxious to go back there, now in the employ of the Argentine government. He proposes to spend a couple of days with us and to go over some of the geological problems of the region, which will be very helpful as he has already spent some time here. We had dinner together and then sat up rather late talking, in Spanish, of course, as I know as little Italian as he does English.

Economic: The presence of the railroad might seem to imply a degree of economic importance and progress which in fact this country is very far from possessing. The railroad itself was built by the government, does not have to make a profit, and doubtless never will. Agriculture is limited to a relatively minute area and does not begin to supply even the very modest local needs. Petroleum is still very unimportant and with no prospects of much expansion. This is a sheep country, and even for sheep it is extremely ill adapted. At present sheep raising does not pay even a bare livelihood. Even with improved prices, this industry could not expand here as the country is already supporting all the sheep it can. It could hardly compete even with southern Patagonia, let alone the fine sheep regions in adjacent areas. The part of the country where we are working is in about the stage of civilization of the North American West in the 1850's, but it seems to be quite devoid of the natural resources that led to the later development of our West. This plateau region of central Patagonia is desperately poor, and so far as I can see it must always remain so.

That does not wholly sadden me. It is good to think that there are some parts of the earth that may be destined never to be populous or civilized. A thoroughly subdued planet would be dull, and it would degenerate. I think mankind needs Patagonia as a permanent outpost, as a sort of museum specimen of the raw earth. I should like to transfer to such a place, a few at a time, those descendants of our pioneers who now complain if they have to leave concrete roads.

Nov. 27.

The last few days were spent with Feruglio, ranging rather far afield and making observations, not without much difficulty for the wind is even worse than usual.

Yesterday we visited the Ramos of Casa Ramos. His house was already well known to us, but he had been away. He was irrigating his garden when we arrived, and we approached too closely in the car and got thoroughly mired. He called out his numerous sons, and with shovels and poles and many exclamations of "La gran siete!" we got out again, had maté with the family (exclusively masculine) and went on our way rejoicing, with a tank full of slightly salty drinking water and a magnificent gift of some lettuce and fresh onions.

Ramos padre is a swarthy, bemustached individual who looks as if he should be holding up lone travelers in the Pyrenees, but who is (like so many tough-looking people) a man of natural charm and worthiness. Still more surprising, he is a visionary, and a very unsuccessful one. Spanish by birth, he came here about twenty-five years ago and has spent his youth and a small fortune in money trying to make the desert blossom. He built a good house and laid out a large garden near the lake, only to have everything completely wiped out by a sudden rise in the water level. Undaunted, he built again, more modestly, and laid out a smaller plot of agricultural land. This was quickly covered with pebbles by a spring freshet. Then he built a dike, miles long to divert the freshets, but this was soon washed out. He built a kiln to burn plaster, and built a very expensive and difficult road up the barranca to haul his product to the railway. The road washed out almost at once, but this did not greatly matter, for he could not sell the plaster anyway. He is still trying, still cheerful, the soul of hospitality. He has not surrendered to Patagonia. He keeps his sanity and his dreams. Perhaps he is really a successful man.

Today has been absurdly varied. At an early hour, we sallied forth for Cerro Blanco, the new field of our activities. When we entered the flats just west of here, we ran head on into a wind which almost makes the previous gales seem like gentle breezes. I do not think that it would be any exaggeration to estimate that it was blowing steadily at one hundred miles an hour near the ground. The air was so full of sand that it was like driving in a thick white fog, with everything blotted out beyond a radius of a few yards. Our previous tracks were obliterated by the drifting sand, and from the crest of each little dune came wisps of flying particles, like clouds blowing past a high mountain peak. Making almost no headway and in imminent danger of getting lost, we turned back and groped our way to Don Francisco's fishing shack. Rain began to come down in sheets, rather unusual when the strong wind is blowing, but there is no end tó Patagonia's nasty versatility.

We drank maté with Don Francisco, the Boy-with-a-horse, and a stranger, an Italian with a Ford. Water fell and wind howled outside, and also to a certain extent inside, for the construction of the shack does not keep the two entirely separate. We talked of revolutions, bones, automobiles, and world economics.

Suddenly the storm ceased, the sun came out, and the wind died down to a normal thirty or forty miles an hour. We started again for Cerro Blanco. Just as we had passed the dark mass of its opposite number, Cerro Negro, we ran out of gasoline, a serious accident here [and one which never

again befell us in Patagonia, I may happily add to this record].
I sent Justino back to camp, with a prospect of a twenty-
mile walk, to bring gasoline, while Coley and I proceeded on
foot to our destination, only about four miles distant, and
put in the rest of the day working. At sunset Justino turned
up with the truck, fortunately having been able to borrow
Don Francisco's flivver for part of the return trip.

Coley drove the flivver back from Cerro Negro, where Justino had left it. When old Fords die they come here. Justino
calls them the "Godfathers of Patagonia." This one is almost unbelievably ancient. It still goes, but it is almost
impossible to steer it and it has a strong tendency to leave the
trail and hunt for a tree to climb—a hopeless ambition here,
however laudable. On the way home it once jumped out of
the track and described a complete circle before Coley could
stop it and ease it gently into the right direction again.

On the way, we ran down and caught alive a chulenguito,
a baby guanaco only a few hours old. It was with its parents,
and they slowed down for it at first, but when we approached
closely they left it to its fate, hung around for a few minutes
yammering, and then wandered slowly away without a backward glance.

In the evening, in the free Patagonian style, we had a
caller, a man with a horse and a dog (not the Man-with-a-
horse-and-a-dog, but another man, another horse, another
dog). He is spending the night with us, but his name, condition, origin, and destination remain unknown. Vera Córdoba also dropped in for the evening, to ask us to take him
to Comodoro the next time we go. He and the wanderer
elicited a brief lecture on geology, asking whether the animals
we find are pre- or post-diluvian, whether the world is over
two thousand years old, whether there have been more than
forty generations of men, and so on. I replied that I had
thought about these things for a number of years and that for
my own part I believed that all these animals are extremely
antediluvian, that the world is many millions of years old,
and that there have probably been tens of thousands of gen-

erations of men, or of the ancestors of man, but that everyone must have his own opinion and that I would not venture to say that mine was right. To my surprise, after this almost too elaborate effort not to tread on their probable religious principles, they agreed heartily. They were sure, for instance, that the fossil wood so common here could not turn to stone in less than millions of years.

This is an acute inference of a sort which might be expected from men who live close to nature but which in fact is very unusual for them. Acuteness of physical observation is, I think, generally linked in such cases with peculiarly dull metaphysical inference. Men who pass their lives out of doors commonly have a vast store of objective knowledge, but their comprehension of any real interpretations of these facts beyond the strictly visible horizon is usually ludicrously scanty. My visitors of this evening seem, then, a cut above most men of their sort whom I have met.

Comodoro Rivadavia
Nov. 30.

Yesterday we finished work from Base Camp One, and today we drove to Comodoro. We called early for Córdoba and found him surrounded by his wife and children, of whom there are now five and will soon be six. He is a burdened and a serious man. Of his eight or nine hundred sheep, a few have been placed in his wife's ownership, but the others are heavily mortgaged and the bank is on the verge of selling them.

"I never drink or smoke," said he. "I used to smoke, but I gave it up. It is very bad. Bad for the health and for morals, I mean. I could smoke, of course. Do not think that I cannot afford to. Oh, no. I have the money to do it, all right. But I gave it up because it is bad. So is drinking. It is very expensive, but I gave it up because it is bad. You should not drink or smoke either. Now you see what a family I have, and how well I get along without these things, I hope you will give them up too."

We assured him that we were entirely convinced that he could afford any minor vices in the calendar and gave them up from sheer will power, and we promised to think the whole matter over very seriously.

He wanted to pay us when we had brought him to Comodoro, and it was difficult to refuse. He is so afraid that we will think he is poor, just because he lives in a one-room hovel with a wife, five children, and nearly another, and can barely achieve the most meager necessities of life. Then too, he is a little shamed before us for another reason. He charged us eight pesos (now less than two dollars and a half) for a large sheep. We thought this expensive, so asked the manager of the bank in Comodoro what we should pay. He said that he really knew nothing about it, but thought eight pesos was quite a bargain. Later we found that this is over twice the market price, that the bank has excessive mortgages on these particular sheep, and that this manager had told Córdoba to put this outrageous price on them for us. I think his part of this deception weighs heavily on Córdoba's mind, and I only wish it weighed even more heavily on the bank manager, but I fear it does not.

CHAPTER VI

HOW TO BEHAVE IN PATAGONIA

Iᴛ is a popular belief among those who dwell in cities, those who consider themselves as sophisticates and exponents of the complex life, that the inhabitants of the Great Open Spaces are simple people whose lives are not patterned or hedged about by conventions. When harassed by modern life, we sometimes sigh for the supposed directness and simplicity of the pioneer or the savage state of society.

To anyone who has really lived among natives of uncivilized communities or who has spent much time under frontier conditions this legend is laughable. Nowhere are beliefs more stereotyped and behavior more circumscribed than among such people and in such places. And nowhere are the penalties for transgression of the set rules heavier. What would merely be interesting eccentricity in civilized life may be literally punished by death on the frontier. The standards are usually very different from those familiar to us, but they exist and are rigidly enforced.

Even in the minor matter of the formulas of politeness this is very striking. With us, for instance, strangers must ordinarily be introduced. The usual absence of this custom in wilder regions is not a simplification. On the contrary, this convention is there commonly replaced by the still more stringent rule that one must not inquire into the name or business of a stranger.

Patagonia has its own etiquette. Some allowance is made for the ignorance of foreigners, but the traveler who does not wish to be treated with a mixture of suspicion and scorn must early acquire the rudiments of Patagonian politeness.

In Buenos Aires and even in the fringes of its culture, as among the more cultivated residents of Comodoro, the conventions are mainly European, but out beyond the Pampa de Castillo things are rather different.

A stranger arrives. His name or business must not be asked, nor must he ask yours, unless there is some very good reason for doing so. He must be invited into the house and must at once be offered maté. If he chooses to stay, he must be fed and given such shelter as is available until he decides to go. Unless he means to stop for some time he will not ordinarily take off his hat, except to eat, and not always then. If a passerby does not actually call attention to himself, no notice whatever should be taken of him. You may stop at a house, draw water from its well or spring, and rest in its shade, and the occupants will go on about their business as if you did not exist, unless you speak to them, whereupon they will be obliged to adopt you into the family temporarily.

In short, the golden rule of contacts with strangers in Patagonia is to mind your own business and to let them mind theirs, to proffer hospitality freely and without question, but only after some sign that it is wanted.

Thanks are never given in anticipation of a favor. To reply "Gracias" to an offer is to refuse it. When being served, "Gracias" is equivalent to "Enough."

When entering a room where anyone is eating or drinking, one must say "Buen provecho!"—"May it do you good!" All must reply "Gracias!" The same formula must be used by anyone who leaves a table before the others are through. To leave a table or a room one may also say "Con permiso"—"With permission." The answer is more often assumed than given, but if given is "Es suyo"—"It is yours."

In Buenos Aires "Sírvase" is a usual phrase for "Please," but in Patagonia it has only its literal meaning, "Help or serve yourself." "Por favor" is the formula for "Please" but is seldom used. "Gracias" means "Thanks" of course, as it does the Spanish-speaking world over, and "De nada"—"For nothing"—is "You're welcome."

"Buen día," "Good day," is the greeting until after lunch, and then "Buenas tardes," "Good afternoon, or Good evening." Saying goodbye has many more subtleties than in English. Aside from the usual good wishes, such as "Que le vaya muy bien," "May it go well with you," the phrase depends on the expectation and desire to meet again. If there is to be another meeting during the same day, say "Hasta luego"—"Until soon." If on another but fixed day, "Hasta mañana," "Hasta domingo," etc.—"Until tomorrow," "Until Sunday." If the expectation of meeting is indefinite, "Hasta otro día" or "Hasta otro momento"—"Until another day, or moment." This is the usual farewell to travelers, with its polite implication that one does not wish definitely to say goodbye but wishes to see them again, even though this may be obviously impossible. "Adiós," contrary to our expectation, is used only very rarely. It is too strong a word. It means that the parting is really very serious and that one expects or fears that it is final. It might be used to confide the care of loved ones to God when they are going on a long and dangerous journey. To use it for a more casual parting may even be insulting, with some hint that one does not wish to meet again.

On meeting a friend, "Qué tal?" "How goes it?", is colloquial, and the answer is "Bien, gracias, y Usted?", "Well, thanks, and you?" This reply is so automatic that it often precedes the question. Several times when I forgetfully neglected to inquire "Qué tal?" my friends said "Bien, gracias" anyway, as if I had. Good wishes are returned by saying "Igualmente"—"Equally," or "The same to you!"

In English we illogically ask pardon for doing a thing before we have done it. Not so my Patagonian friends. They may ask for permission to do it, but never ask pardon before it is done, and very seldom after.

"Salud!", "Health!" carries a large burden of Patagonian politeness. It is a common greeting and its answer. It is an obligatory wish, for which thanks must be given, when anyone sneezes, for the curiously widespread folk belief that there is some spiritual danger in sneezing occurs there too. In drinking wine, or any other alcoholic beverage, even during a meal, the drinker says "Salud!" If the others are also drinking, they repeat the word. If not, they say "Gracias!"

These, and many more, are the formulas of politeness in Patagonia, and their omission stamps one as a boor even more surely than would a similar slip in the best circles of our great cities. I do not mean to say that Patagonians are universally and equally polite by their code. There are boors in Patagonia. On the whole, however, I think there are fewer proportionately than in most civilized places.

The real center of Patagonian social life—perhaps of life there in general—is maté. If I were only more lyric I would pen an ode to this wonderful beverage. I can at least explain, somewhat belatedly, what it is.

"Maté" is an Indian word (Guaraní, I think) and it means simply "gourd," but it has come to be applied to a drink which used always to be, and still usually is, prepared and served in a gourd. Yerba maté, or more briefly yerba, is its basis, the prepared leaves of a tree, *Ilex paraguayensis*, related to our holly. These trees, tropical to subtropical in habit, originally grew wild in Paraguay and adjacent regions, where the Indians drank maté long before the coming of white men. With the spread of this custom, the wild stock no longer sufficed and there are now many great plantations, called *yerbales*, where yerba is grown and prepared. This has become a great industry and in the maté drinking countries it is quite as important as is the coffee business in the United States. There are many different brands and grades of yerba, and in the cities large billboards flaunt the claims of rival sorts and the newspapers and magazines carry innumerable advertisements for them.

The finest yerba still comes from Paraguay, but there are also very extensive yerbales in the northern Argentine and in southwestern Brazil. Border wars or skirmishes have been fought over the territories where the best trees grow. Preparation for market is simple. Leaves are stripped from the trees, together with twigs and some smaller branches, and these are dried or slightly roasted and then ground. In the old days, hutlike frameworks of lattice were constructed, the leaves and branches were laid thickly over these, and fires were kept burning underneath for the necessary time. When crisp, the leaves were sewn into bundles covered with fresh cowhide. As the hides dried, they shrank and compressed the yerba until it was a mass almost as hard as iron, compact and convenient for shipping. Now on the more modern yerbales the roasting is done in rotating iron ovens and the yerba is then shipped to large central mills where it is ground and packed.

Some yerba consists purely of leaves, with even the stems removed, while other brands and qualities have the stems and often also small twigs. These are not entirely adulterant. They give a distinctly different flavor, more pleasing to some palates, and they also help somewhat to keep the mass from clogging the bombilla, described below. The final product looks like just what it is—ground leaves. Chopped alfalfa looks much the same, although the yerba leaves are originally considerably larger than those of alfalfa. As it leaves the mill, the yerba is packed in small tins, of one-half or one kilo usually, or else is compressed very tightly into round cylinders, with burlap sides and wooden disks at the ends, weighing usually from five to fifty kilos (eleven to one hundred ten pounds). Great carts, like brewers' wagons, loaded with these cylinders are a common sight on the streets of Buenos Aires. The price of the usual sort runs from about seventy centavos to a peso and a half the kilo, roughly fifteen to thirty cents a pound (with the peso at par). The difference in price depends largely on delicacy of flavor and on strength obtainable from a given amount of yerba, which is highly variable, and these in turn depend chiefly on the particular trees from which the yerba comes, and where they grew.

There are several ways of preparing the drink, maté, but of these the original method and, to my taste at least, the best is that of making *maté amargo*. A small, hollow, dried gourd, open at one end, is used. This is the maté, strictly speaking, from which the drink derives its name. Matés may be very plain and simple, or they may be highly decorated with incised pictures, designs, and emblems, often painted and inlaid with gold leaf. Dry yerba is placed in the maté until it is about half full, or a little more, and enough cold water is poured in to make a thick paste. When this is well soaked in, hot water is added.

Drinking is done through a metal tube, the *bombilla* (pronounced bome-bee'-zhah in the Argentine), with a mouthpiece at one end and an enlargement at the other end which is pierced by many small holes and acts as a strainer. The details vary a good deal, from the simplest possible tin tube, to elaborate affairs of solid gold and silver carved and hammered into designs and with an intricate strainer which can be taken apart for cleaning.

The brew, when the maté has been filled with hot water for the first time, is not good. It is intensely bitter and the first drinker also sucks up all the fine particles which can pass through the strainer of the bombilla. It is therefore very bad form to offer the first drink to anyone. In urban society the server must drink it himself, as best he can. In Patagonia, it is *de rigueur* for the server to draw this into his mouth and then to spit it out explosively onto the floor— the floors are all of dirt and this merely helps to keep them tamped down and hard. The yerba is left in, the maté is filled with water again, and is ready now to be drunk with full enjoyment. This can be repeated many times, fifteen to twenty if the yerba is good and is properly served.

The water used is at first only warm or barely hot, and for successive servings it is made hotter and hotter until it is boiling. In this way the strength of each serving is kept about the same, hotter water being used as the yerba becomes exhausted. To accomplish this, there must be two kettles, one of cold and one of hot water, and the server should be next to the fire so that he can keep up the temperature of the hot water.

Making maté properly is an art. They say that a good maté maker, *cebador*, must be born with a talent and must cultivate it assiduously. The process looks and sounds simple enough, but there is great judgment required in using exactly the right amount of yerba, soaking it enough and not too much at the start, and keeping the temperature of the water exactly right throughout. To a developed palate, the differences in flavor due to the slightest variation in these details are quite apparent. I know that I have never learned to make maté quite so well as Justino, who added this ability to his many others. And Justino, with modesty very, very surprising in him, admitted that he was not a top-notch cebador.

There are other things that must be known in order to make maté amargo properly. For instance, a new gourd must never be used. When one must be purchased, it is filled with damp yerba of good quality and allowed to stand for several days, renewing the water from time to time and taking care that it does not ferment. The inside of the gourd is slightly porous, and it soaks up the essence and becomes seasoned, like an old pipe. This makes a surprising difference in flavor, apparent even to an untrained taste. The older a gourd is, the better, unless through carelessness it is allowed to ferment or to mould, which spoils it forever. The natives are so reluctant to throw away old seasoned gourds that they will go to almost any lengths to repair cracks or breaks. The way to repair a crack is to take part of a sheep's intestine that is shaped like a little bag, turn it inside out, and put the gourd inside while the intestine is still wet. This shrinks on drying, grasping and compressing the gourd, and becomes hard and leathery. It has, furthermore, an attractive honeycomb-like pattern.

Drinking scalding-hot maté through a metal tube takes some practice, but at the expense of blistered lips we finally acquired the knack of doing it without discomfort.

Maté is drunk on every possible occasion, and with no occasion at all. The true criollo drinks it almost constantly, fifty to a hundred gourdfuls a day not being considered excessive, although my gringo appetite is well satisfied with about twenty a day. This is not, however, as much as it might seem, for the yerba takes up so much space that a gourdful is comparatively little. It would take perhaps five to be equal to a teacupful of pure liquid. Baliña assured me that it had been many years since he had drunk any water, maté supplying all his liquid needs.

Although its use is thus often solitary, maté is at its best in its social aspect. Then is when it gives real peace of soul and communion of spirit. At any gathering in Patagonia, whatever the occasion or the time of day, the first thing done is to serve maté, and there is a little ritual in this.

The company forms a circle, with the cebador next to the fire. He prepares the yerba, brews and spits out the first gourdful, and then brews the second—and, by the way, no time should be allowed for brewing; it should be drunk immediately after pouring water or it will become too bitter and eventually will turn black. The second maté is passed to the man on the left of the cebador, who drinks it and passes back the gourd, and so on around the circle in rotation. Only one gourd and one bombilla are used, no matter how large the company. This is neither sanitary nor enticing, but one becomes accustomed to anything and either to refuse maté or to insist on a separate bombilla would be insulting.

As long as a slight foam forms on the surface when water is poured in, the yerba has not lost its savor. When this disappears, the cebador knows that he must dump out the yerba and start anew. Before serving each time he pours a little hot water onto the index finger of his left hand to test its temperature and be sure that he is achieving perfection. When anyone has had enough, he says "Gracias!" when he hands the gourd back to the cebador, who must reply "Buen provecho!" The circle continues until all have said "Gracias!"

The flavor of maté is slightly bitter. The only comparison I can make is with green tea, but this comparison is not close. Now that I am accustomed to it, I definitely prefer maté to tea of equal quality. Relatively few people like maté the first time they try it, but I am convinced that the same is true of tea or coffee. It is probably because of this necessity to acquire the taste that maté has never become popular in North America or in Europe. Attempts have been made, and are now being made, to introduce it here, but with very little success. Yet it is so firmly entrenched in South America that it ranks with tea and coffee as one of the great non-alcoholic beverages of the world. Like tea, it is a cup that cheers but not inebriates, and unlike tea it contains no tannin and seems to have no bad effects. It does contain mattein, a compound closely analogous to thein and caffein, to which it owes its definite but mild stimulating effect. This effect is wholly pleasant. It is bracing. It sets all right with the world. Coming into camp tired and bedraggled, gloomy and at odds with myself after a hard day's work, I begin to cheer up after the first maté, and soon decide that life is worth living after all.

The real reason for drinking maté is liking it, its flavor and its beneficent effect, but it also seems to be good physiologically. Many people in the Argentine have almost literally nothing from one year to the next but meat and maté, and yet they thrive on this apparently unlivable diet.

Besides maté amargo there is maté cocido, brewed in a pot, strained, and drunk from a cup with cream and sugar. With this effete and citified innovation I will have nothing to do.

It is curious that one of the few native Indian drinks of North America, the Black Drink of the southeastern tribes, was similar to maté and was also made from ilex leaves, but of a different species. Its use has now wholly died out, and deservedly so, I think, for I recently got some of the leaves and prepared some Black Drink, which I found not only vastly inferior to maté but even positively distasteful.

Happy are the Patagonians to have their maté! They have so little else. Here is life stripped almost to its barest essentials. Many tribes of savages living in the depths of jungles have more in the way of material comforts than has the average poor puestero, or small sheep rancher, of central Patagonia. His material wealth consists of a hut, a few leagues of barren ground, and a flock of sheep. Even these things are his only by sufferance. Usually he has only grazing rights to the land, and not true ownership, and the house, even though built with his own hands, belongs to the land and not to him. The wool crop from his flock is always mortgaged far in advance, and usually the animals themselves are mortgaged and remain in his hands only as long as he can make their products pay interest, a desperate, almost hopeless struggle under present conditions. None of these nominal possessions can be sold. He sees no money from one year to the next. At best he manages to keep up a small credit at one of the stores.

From this credit, usually near or at the vanishing point, he must purchase all the necessities which he cannot produce himself, and also what few luxuries he may have. He has mutton in abundance, and this is his chief and not infrequently his only food. Yerba is a necessity and must be purchased. If credit permits, he will also obtain flour, potatoes, and tobacco. Wine is definitely a luxury, and while almost all would like to have it, in recent years few have been able to afford it. Clothing, selected entirely for cheapness and durability, must also be purchased. For his horse, unless abjectly poor, he will buy a carpincho skin and make everything else himself.

These are the sole purchases of the average puestero. Beyond them are the few possible frills of life, sweets and extra food, "town clothes," cartridges, phonograph records. Most of the country people have firearms, but they are very seldom used. Cartridges are too expensive. Covetous glances were frequent, but the only thing I remember any of these poor but proud people definitely asking for was ammunition. Three of the puestos that we visited had phonographs, ancient, antediluvian contraptions that had finally drifted to this end of the earth. A phonograph is wealth here, the badge of material success. Outside of the towns and one country inn, we saw no radios. There are a good many old motor cars in the country, but they belong to the carriers and the townspeople. No puestero has one.

Many of these people can read, an ability more often passed on from father to son than learned in any school, but reading matter is excessively scarce. Most people have managed to acquire a few old newspapers and magazines, but books are almost unknown. There is no intellectual life, no diversion, almost no contact with the outer world. News, even that of vital local import, travels slowly, from mouth to mouth.

In summer, there are the sheep to attend to, lambing, dipping, shearing. Apart from these set occasions, the flocks receive very little attention. Except for the neighborly asado, the one social function of the country, and for occasional wayfarers, life proceeds on an even tenor, a stream in which one day imperceptibly follows another and all flow on ceaselessly toward a lonely death. When winter closes down, the doors are battened shut, and only brief sallies to see to the sheep break into silent weeks spent in the gloom, drinking maté and sleeping.

Life is encompassed by the darkness of the hut, the solace of maté, and by a barren outer world where the ruthless, the nerve-racking, the terrible wind blows incessantly. Some go insane. I cannot erase a horrible picture from my mind— two men in straitjackets lying neglected on the beach, writhing and shrieking. Later they were loaded like two logs onto the boat for Buenos Aires. They escaped from Patagonia. Their dream of leaving this desolation and of going at last to the fair and bright city came true. The sane do not escape. Others become morose, then bitter, then brutish. They go through human motions, but their lives are those of animals.

Patagonia seems sometimes to be a personified force, an evil and malignant spirit, delighting in the torture of souls, seeking with unfair weapons a crushing victory over mankind. Yet there is more than this. As in listening to a symphony, as before a great painting, or as in the words of a poem, there is often in the midst of this misery a glimpse of the grandeur of human life. Man is the animal that is above all animals. Something in him is greater than the sum of his days or the result of his actions. After its worst effort, Patagonia falls back abashed before this, and I think it reserves a savage love and an intimate delight for those strong enough to defy it.

There is laughter in Patagonia. There are those, there are indeed an amazing number, who achieve the great victory of being commonplace there. The normal men of that fierce country are geniuses in living. Their strength is the rare strength of insensitiveness, of not being bored with blankness, of being unable to perceive horror. It is stupid, if you insist, but stupid only by a standard which would fail utterly in Patagonia. It is the brilliant men, by such a standard, who are weak and who escape into insanity. The strength of the others, the victors, is wholly admirable, and only the more so because it is so completely unconscious and so entirely devoid of introspection.

These puesteros are the true Patagonians. It is through and for them that the country lives. They inhabit the pampas. They are not a race, but they are moulded into something of a common semblance by the land. Their forefathers came from every corner of Europe, and often by devious routes. In not a few of them, too, there flow some drops of aboriginal blood, something from the tall Tehuelches and from the fierce, unconquered Araucanians. Now they are simply Patagonians, a word no more definite in ethnic meaning than is "American," but one with a real connotation nevertheless.

Even here in central Patagonia, these people are in a minority. They occupy the land, and without them it would die, but the others outnumber them.

There are the townspeople. Of their life I have said something and will say more. They do not differ very greatly from tradesmen, innkeepers, officials, and hangers-on the world over. Then, bridging to some extent the gap between the people of town and of country, there are the carters. Sometimes with motor trucks, sometimes with carts, sometimes, but more rarely, simply with pack animals, they tie the country together. They are its circulatory system, hauling out the produce of the country and bringing in the imports of the coast towns.

Beyond this, as in all societies, there is another and a sparser tribe: the wanderers, the individualists, the hunters, and the outlaws. The nearest thorn bush is their home. Their possessions are a horse, a knife, boleadoras, and the clothes they wear. They live and they die by their wits. They accept hospitality as their right, but never beg it as a favor. They sleep wherever they are and they eat whatever they can get. If, as frequently happens, they have the misfortune to kill someone or to steal too much, they wander on, traveling at night, north or south or westward across the cordillera as fancy strikes them. They are not, as a rule, evil at heart, but they take what they want and they kill when they think it necessary. They do not recognize rules or laws. If they did, they would not be in Patagonia. No one knows them, and they know no one.

How many of these solitaries there are, it is impossible to guess. They may pass through a wide region without ever being seen. Central Patagonia remains a stronghold for such nameless, homeless, friendless men. What dark secret drove them here, or whether that secret lies only in some inner urge of their forlorn souls, cannot be told. They do not talk of themselves. At work out in the wilderness we more than once saw a strange rider silently passing near us who would merely turn his head away and increase his horse's gait when hailed.

Some of these men are hunters, not merely for themselves but with some thought of profit, and these may have a more circumscribed range, although they seldom have fixed habitations. It is here that a classification of Patagonian society impinges again on the puesteros, for some men alternate or combine hunting and sheepherding. They hunt chulengos, skunks, foxes, and ostriches. The hides are made into quillangos, fur blankets, locally, or they drift into world trade and show up again in New York and Paris under strange names.

The ostriches are hunted for their feathers. They have no plumes, but the feathers have a small value for feather dusters. In Comodoro we saw dusters made of feathers that had been shipped from Comodoro to Germany, there simply fastened to sticks, and then shipped back again.

There are other classes of people, but I do not feel that they belong on this canvas. There are, for instance, the oil field workers and officials, but they are a thing apart from the true Patagonian scene. The oil is an accident here, and it is not permanent. And there are farmers in Patagonia, but they, too, can hardly claim a significant part in this picture. Their foothold is so meager—in the great bulk of Patagonia, from the Río Colorado to the Straits of Magellan, they have only a few narrow strips of land, at best a mile or two across and to be numbered on the fingers of two hands. With more right to a place among the Patagonians are the estancieros, the owners of the great sheep ranches and their personnel. Here in central Patagonia, the most desolate and barren part of the country, there are almost no big estancias. These typify rather the country both to the north and to the south, and I shall have some occasion to speak of them when our travels have taken us farther afield.

Even in Patagonia life changes and social values shift. If the sheep ranchers hold firm (which, however, is not a certainty), then the nomads must disappear in time. Hunting as a profession is already rapidly passing. The animals are becoming less and less abundant. While we were there the hunting of chulengos, formerly a mainstay of the country, was prohibited. Conditions of life at the puestos will, I hope, improve, although any improvement will be slow. Communication is becoming easier and more rapid. Roads are being laid out and will probably be built and improved. Something more of world standardization will creep in as isolation decreases. But the real characters of this country are immutable: wind, cold, drought, pebbles where soil should be and thorn bushes for grass. These cannot change within a millennium, and as long as they remain, Patagonia will be its savage self and its people will be set apart from all others.

CHAPTER VII

HOW NOT TO BEHAVE IN PATAGONIA

THERE was once an estanciero who lived near Puerto Deseado and who had a ranch hand who may be called José. This José was a pretty good sort, take him all in all, but he had his bad moments. One spring day he happened to have a bad moment just at the time when he also happened to have a knife in his hand and when someone who had once annoyed him was nearby.

After the funeral, the Majesty of the Law reached out from Deseado and collared poor José in spite of his kind heart. He was locked up in a cell and given a long time to get over his bad moments while he awaited trial. Out on the estancia the sheep were being sheared, but somehow things did not go well. The estanciero began to miss José very much —I almost forgot to say that in addition to having a kind heart José was a first-class sheep man and that from Carmen de Patagones down to Magallanes there was not his equal in shearing; he had a touch around the ears and tail that was a real pleasure to watch.

So the forlorn estanciero went in to Puerto Deseado and went to see the Majesty of the Law.

"Do you realize," he said, "that here I am right in the middle of spring shearing and that you have the very best sheep shearer in all Santa Cruz Gobernación in your lousy lock-up?"

"Well," said the Majesty of the Law, "I can quite see your point of view, but you know there was that little unpleasantness about the knife, and the widow seems to feel that I should do something about it."

"Let's put it this way, then. Suppose you turn José over to me temporarily. As soon as we are through shearing I will send him back to you."

Thus it was agreed, and José worked long and faithfully, with some particularly fancy work around the tails and no bad moments. When the work was all done, the estanciero said to him:

"Now, José, here is all the money that I owe you. It is a great deal and I am sorry to part with it, but after all you have earned it. And now they are waiting for you in Puerto Deseado, where the Majesty of the Law is planning to hang you. I promised him that you would be back, so just give me your word that you will report at the jail there, and then skip along like a good fellow. Goodbye!"

So of course José went to Puerto Deseado and reported at the jail where he was immediately locked up, he and all the money he had earned shearing sheep so well. The next day he asked to see the Majesty of the Law.

On the second day, José turned up healthy but broke at the estancia, where he asked for work and was given it.

"Oh, well," the Majesty of the Law was saying at about the same time, "it was just a bad moment after all, poor chap." And he paid cash for a very shiny new automobile.

That is a true story, and here is another:

It was two o'clock in the morning and Comodoro Rivadavia slumbered peacefully under martial law and the Southern Cross.' It was a holiday night, but from only one house still came sounds of subdued merriment. For convenience I will call the house that of Jack Davies, and for further convenience I will call two of the guests Coley Williams and me.

Even there, the party was over and everyone came out into the patio to say good night. One of the guests, the one that I am calling Coley for the purposes of this story, decided that only one thing was lacking to make the evening complete, so he picked up a shotgun which he happened to have handy and fired both barrels in rapid succession.

The noise was terrific and the results were instantaneous.

Women shrieked. Men yelled. Windows flew up. Babies howled. Whistles were blown.

In New York very few people would turn their heads if a shotgun were fired off at two A.M., but that is one of several differences between New York and Comodoro Rivadavia. This was not long after the revolution, and it was in the midst of many rumors of new revolutions and counter-revolutions. Nor has Patagonia forgotten the terrible revolt of the peons when so many people were slaughtered without warning. A shot in the night there means murder at the least, and is more likely to be a signal for a massacre or a revolution.

We, as I am calling my protagonists, departed immediately in our car. The streets were full of running people. The old-fashioned night-shirt is still in high favor in Patagonia—the whole incident seems quite worth while if only because it permitted that observation, which could otherwise have been made only after very prolonged and difficult investigation. At the first corner we came upon an extremely agitated little Indian policeman.

"Have you just come in from the country? Did you hear a shot? Where is it?"

"Perhaps a cold motor backfiring?" we replied noncommittally.

The policeman persisted in his very natural error that we had just come in from outside town, and we were too excited to correct him. He waved us on. We went to the Hotel Colón where we were soon asleep, and so slept on peacefully until morning.

In the meantime the source of the noise had been fairly well determined, and waiting only for strong reinforcements the police moved down in a body to the Davies house. While they pounded on the front door, Davies, his wife, and their remaining guests slipped out the unguarded back way and retired in good order to a bar at the opposite end of town.

They reached there only a few seconds ahead of the now thoroughly alarmed police force. In spite of indignant denials of all knowledge of the affair and of remarks on the rights of Englishmen, the whole party was lined up against the wall and searched for weapons. The officer in charge then explained that this was a capital offense, and that he intended to haul Davies off to jail for it. So Davies broke down and said, yes, the shots had been fired from his patio but they had been fired by one Williams, now presumably either in flight or at the Hotel Colón.

Then the police galloped over to the hotel and routed out Bruzio. Did he have a guest named Williams? He wasn't quite sure, why did they ask? Because they meant to arrest him at once! Why, yes, now they mentioned it Bruzio did have a guest named Williams, in fact he had two, and which did they want? They didn't know, better produce them both. But no, señores, that Bruzio could not do, for then obviously he would be injuring an innocent man. One Williams must be innocent, so how could he wake them both up and subject them to the shock of police scrutiny? This flawless logic baffled the police altogether and they retired to consult with their superiors.

[You are a liar, Bruzio. You had only one Williams there, but I think you will be forgiven in Heaven. You are also a bandit, as I knew each time I paid your bill, but I forgive you that, because you gave us service when it mattered.]

Meanwhile we slept on, quite unaware that others were suffering for our sake. My first intimation of disaster was when I came down the next morning and found the whole hotel seething with swarthy policemen, and Davies in very shattered condition clutching a gin-and-tonic to his bosom. He soon explained the affair and its very real seriousness.

It was a fact that Coley legally could and should have faced a firing squad for that moment of hilarity. The announcements of martial law then plastered all over town carefully specified this penalty for his exact offense. Yet there was nothing for him to do but to give himself up.

Davies had just extricated another hapless American from jail and he felt that his influence was temporarily exhausted, so it was to the bank manager that we turned in our need. Davies finally repented and went too. When he and I reached the Comisaría of police, there was Coley, with the range of the place but technically in durance vile, and there was the bank manager, warming a chair. He looked very, very unhappy. One could see him squirm as he thought of his wife's remarks about him and his rowdy friends, but there he was.

The comisario quickly dropped talk of martial law, and began to harp rather heavily on ten days in jail. Our emissaries explained that this was impossible, as we were leaving for the field immediately and Coley was needed there. The comisario came out to see me, where I waited apprehensively in the outer office. I was in field clothes, as we had planned to leave immediately on rising that morning, and I had a luxurious but unflattering red beard. He looked me up and down, clearly quite the opposite of impressed with my importance, but finally he murmured, "Ah, very intelligent, very intelligent!" I manfully resisted the impulse to ask for this in writing.

After much more talk, it all boiled down to fifty pesos fine and confiscation of the weapon. Now this shotgun had been especially made for Coley and was his dearest possession. Its loss would have been a blow almost too great to be borne. With the connivance of one of our friends, I dashed out, hid Coley's gun under my mattress, and then went and rented another gun from the local jeweler-pawnbroker. This was an old, very rusty 16-gauge weapon which was wholly beyond its useful days. It would surely have exploded and maimed anyone rash enough to try to shoot it, but at least it had the outward aspect of a shotgun. While Coley was washing his hands after the indignity of being fingerprinted I managed to whisper to him a description of the gun with which he was now provided.

In a procession we all went to see the judge. After some more talk and after the quiet transfer of a moderate sum to an underling who presumably knew where it should go, all was arranged. When told that this was the very weapon that had caused the great riot, the judge looked very skeptical, as well he might, but he said nothing and wisely decided that the gun was not worth keeping, thus saving us our deposit with the proud owner. To forestall criticism, a statement was drawn up and signed by Coley. This masterpiece of humor went about as follows:

"The undersigned deposes and says that on the specified night at the hour of two A.M. he discovered that his shotgun needed cleaning. Finding himself then in the patio of Señor Davies, he proceeded to clean the gun in that place. Unknown to him the gun happened to have both barrels loaded. In the process of cleaning, one barrel was accidentally discharged. Proceeding with the cleaning, to his intense surprise and pain the second barrel also went off almost immediately. Both shots happily went into the air without producing the slightest hazard of life or property to any person or persons."

The changeling gun was returned, with the gratuitous statement that only good guns were confiscated. All were happy and we parted with many expressions of mutual esteem. Total cost, a few pesos, a few nervous minutes, and a day— the day hardly wasted, as in it we saw Patagonian justice in action and learned something of its motives and processes.

And that is the second true story of the operation of the law on the frontier. The third is anticlimactic but it is apropos because it introduces a new and excellent character into the cast. Without reservation, we here appear under our true names, except for Scottie whose true name I never learned.

At last we were ready to go off to camp, and our friend Tobin wished to go with us (of his visit, more later). But we were in difficulties, for Tobin must return within a day or two and we could not spare the time to drive him back. Good old Scottie leapt into the breach:

"By gum, a car you want? Take mine, go on, do."

"Oh, no, Scottie, we couldn't take your car. Anything could happen to it way out there across the Pampa."

"Take it, take it, me boys. I won't take no. You must have it. Drive it out past Kilómetro Tres. Go on."

Such generosity from Scottie, who usually very little belies his name! We should have been suspicious, but we succumbed and took the car.

In Kilómetro Tres there was much blowing of police whistles and we were soon surrounded. No explanation was given. Orders were simply to bring that car and its occupants to the local Comisaría. Tobin had been driving, but he had no license and this is a heinous offense, so I volunteered as driver.

At the Comisaría we refused a kindly offer to be lodged in a cell and hung around the courtyard, under guard, for some time, still without the slightest inkling of our crime. Finally I was led in to the comisario (not the one who had rashly guessed that I was intelligent, in Comodoro). I was wearing a beret and my guard, an Indian, told me to take it off. Annoyed, I ignored him, but he snatched it and threw it on the floor. After a searching examination of my history, including the names and ages of all my maiden aunts, the comisario finally broke down and told me that the last time that car came through his jurisdiction it had been speeding and that they wanted to fine the driver. With splendid eloquence I convinced him that I had never driven the car before (and the "before" was superfluous; I had never driven it at all!) The fact that I did not even know the name of the owner of the car was a slight flaw in my tale, but we got past that somehow.

Good old Scottie! He had not been quite sure whether they had the car spotted or not, and to play safe he had decided to let us drive it past the police trap for him!

In three easy lessons, that gives some idea of the strange workings of justice and the law in the far south, where the towns are swarming with police and officials, and where beyond the town limits each man is a law unto himself.

I do not want my Argentine friends, to whose eyes these words may come, to think that I am belittling their country, or poking fun at it. Patagonia has as little to do with the

Argentine as a whole as had, for instance, the Arizona of a generation ago with the United States as a whole. In fact it is not necessary to go back into the past to find strange anomalies and irregularities in the administration of the law in the United States, or in any country where pioneer conditions lie still within the memories of living men. In such respects Patagonia differs from other frontiers where progress is unequal and patchy, like a queer checkerboard, only to the extent of some racial traits, which cannot be judged either as better or as worse than those found in other countries.

Some years ago I was working in a remote and rather desolate part of the Colorado mountains. Each morning I used to see the local sheriff driving out to a mining claim which he was developing. With him rode the one prisoner who was serving a sentence in the jail of the nearest town. The sheriff had given his ward a share in the claim. Both were boarding in the jail, and happily going out to work their property during the day. That could serve as an American translation of some of the processes of justice in Patagonia.

But let us return to Patagonia, and to Scottie. I first saw him in the bar at the Hotel Colón, heavy, red-faced, with thin sandy hair and pale, watery, sagging eyes. He dragged himself in, easing his bad leg along by means of a heavy cane, and he sat down with us and gently cadged a few drinks under the mask of a lot of sage advice to the young and vain regrets from the old. Once we were established as fair game, we saw him only too frequently, and we had from him enough autobiographical data to make very full lives for several active men.

It is all a little vague. His tales are curiously inconsequential, and it is often not at all clear whether a given incident is supposed to have happened in Singapore, San Francisco, Edinburgh, or Cairo.

A few years ago, so he says, a lady novelist cornered him and took copious notes at his dictation. She proposed to write up his life, and perhaps she has done so.

"Of course," said Scottie, "I couldn't really tell her everything about my life. Some of the incidents is a little coarse."

He neglected to add, and for her peace of mind I hope that the lady novelist did not discern, that many of the incidents are not true.

I cannot reproduce his brogue, nor can I be quite as vague as he, having to write definite words here. A few of the less coarse incidents can be strung together in some sort of order:

"Begum, young ones is wild."—This is his twelfth year and the beginning of his wanderings. "There was this ship's captain and there were the police notices on every wall in Edinburgh with me pictures. 'Well,' says he, 'ye can stow away and welcome, but mind that I don't know ye.' So there we were in Greenland."

It is all very casual. Somehow he got into the army, later on, and was a sergeant-major in the 42nd—at least that one point happens to be true, for one of his severer critics checked up on it with the War Office. Then there is something about the Khyber Pass, and then about Egypt.

"I was all through that campaign. All up the Nile. Skirmishin', skirmishin' every day, and the sun shinin'. Then we came to that hill. Begum, the 42nd got the chances, too, that the others hadn't. The Colonel says we will take that hill, so we goes out to be killed. Them dervishes was up top, and shootin' and throwin' stones, and I don't know what. Them that was hit fell down between the sharp, sharp rocks. This was at night, o' course, and we wasn't up top till dawn. Oh, we got up right enough. And there wasn't nothing there.

"I got this leg at that hill. Just two splinters here in the knee. Yes, that and the dagger wound in the hip too."

The scene shifts aimlessly to Hongkong.

"Captain got me that job. China Police. Fifteen days he give me too."

"He gave you fifteen days in the China Police?"

"No, no. That was before. India. There was a bit of fighting in barracks and captain comes in. 'A goddam disorderly house it is,' says he. I don't know. Anyway they had me up next morning, and some says it was murder. But, 'No,' says captain. ' 'Twas a mistake,' so he gives me fifteen days, and then out to the China Police. Teachin' 'em, I was. Too damned interestin' it was."

"So then I suppose you went back to Scotland?"

"They'll hang me when I do. Begum! I'll not be back. That constable was a friend of mine. 'Twenty-four hours we look the other way, and then when we see you we hang you.' That was the time I had a hotel in Edinburgh. He must have hit his head on the kerb. I didn't stop to see. Next day it was 'Twenty-four hours.' And him lying there dead, begum."

He wanders kaleidoscopically for a time. Then he pulls himself up for a moment and makes a glimmering of sense.

"There was this ship, and I made a mistake, and the skipper was for hanging me."

"A fairly serious mistake that must have been!"

"Aye, a bad mistake that was. Bad. So I went over the side and there I was on the beach. Near Rio. Walked down to Santos, I did. A long walk. I was in rags. And them nasty shellfish. Aye, that was bad trouble, begum."

Somewhere else—or can it have been the same time?—he had to swim ashore with someone after him. The pursuer caught his shoe, but Scottie kicked it off and reached the shore.

He came to Patagonia, so he says, in a gold rush. There was no gold but he stayed. And there he will stay, because he maintains that it is the only country on earth that has not yet kicked him out. And now he is too old for any really first-class sin and deviltry.

Scottie claims to have been everywhere, everywhere on earth, except in the "central of Peru." He treasures that one little omission in his wanderings. It is the flaw that makes perfection, like a beauty spot on an otherwise perfect skin.

One of the Rivadavians, rather unkindly I think, determined to trip Scottie up.

"So you've been everywhere?"

"Aye, everywhere but the central of Peru."

"Up the Amazon, I suppose?"

"Aye, way up."

"As far as Pará?"

"Far beyond that."

"To Manaos?"

"Oh, aye, far beyond, far."

"To —— ?" (naming a place in the central part of Peru).

"Aye, 'twas there I saved the lad's life. Farther yet we went."

"Well, how far?"

"Way, way up. The headwaters and beyond. I don't just remember the name of the place."

"Was it Pernambuco?"

"Begum!"—slapping the table, a favorite gesture. "That's it. Pernambuco!"

"But look here at this map. Pernambuco is nowhere near the Amazon."

"Aye, so it isn't, so it isn't. Must have been some other place. I see ye're drinkin' whiskey, sir. I'll just ha'e one with ye for old times' sake."

One of our neighbors, Duvenage, who will reappear later in these pages, felt it his duty to warn us that Scottie was not a flawless mirror of the truth.

"Take that leg of his," said Duvenage, "how did he say he got that?"

"He told us once that he got it at the Khyber Pass, but lately he has been sticking to the tale that he got it chasing the mad mahdi in Africa."

"Now that just shows. The old liar just wants to make himself look romantic. I'll tell you how he really got his lame leg. It wasn't off in any of those wild places like India or Egypt. It was right here in Patagonia he got it.

"He had a bit of a row in town, one winter night, and someone got killed. Don't know whether Scottie did it, but he had a bad name so they went looking for him. He cleared out on horseback, heading for the cordillera. It was the dead of winter and a blizzard blew up. He worked the poor horse until it dropped dead in its tracks.

"That horse dropped dead, and Scottie's leg was under it. He was weak, too, and couldn't pull his leg out. The horse froze stiff as a board on top of him. There he lay for two days. Some wandering Indians found him, finally, more dead than alive. They nursed and kept him through the winter, but he never quite got over it.

"And that's how Scottie really became lame, right here in Patagonia. You see, there isn't really anything romantic about the old liar."

When we first arrived in Comodoro, Scottie had some sort of job, vague like everything about him, with one of the oil companies, but he must have made one of his famous mistakes, for he soon severed his connection with the firm, with considerable hard feeling on both sides. He drifted back into town and took up residence in one of the less elegant hotels, which is to say, a hotel not elegant at all. There was a girl there, apparently largely Indian and a pretty little piece in a wild way, who waited on him hand and foot and followed him about like a dog except when he was cadging drinks from visiting foreigners. He earned his living, vaguely of course, by petty trading and anything else that came his way. He used to try to sell us hams and cheeses sent down from the Chubut and the provinces.

There was also an American, one of the few actually resident in Patagonia. Call him El Rey, although that is not his real name or nickname. It suits one phase of his character, and I want to go back to Patagonia, and he is too good a shot for me to feel safe in making his designation too explicit.

El Rey has an enormous area of grazing land back in the cordillera. He came down from Montana a generation ago, determined to make a quick killing and to return to the States to finish his life in luxury. It is an old story in Patagonia. While he was making money, it seemed inexhaustible and he spent it like water, putting off his departure from year to year. Now he is no longer making money and cannot leave. He will never leave.

On his estancia, El Rey is king indeed. It is so remote that nothing can touch him. He is the law and the judge and the executioner. Those who have offended him there are dead, and there is no one to call him to task. He is justified, for life is not gentle in such places (still white on official maps), and if a man does not defend himself no one else will. He has a mark set up some paces away from his door. Every morning when he comes out, he stands in the door and empties a six-shooter at the mark. If one shot misses, he goes back to bed until the next day.

But even El Rey is not fearless. He has a wife, a native, and her tongue makes him tremble as no physical danger ever could.

When I first saw him, it was seven o'clock in the morning and he was sitting at a table in the Hotel Colón in Comodoro drinking Canadian rye whiskey. Bruzio imported it especially for El Rey by the case—no one else in Patagonia drinks rye —and sold it to him by the glass. I said "Good morning," and he poured himself a drink and said, "We might as well be drunk as the way we are!"—a stock phrase with him, especially when as drunk as he could possibly be, I later learned.

He had come down to sell his wool, and his wife had told him the exact price that he must get for it. When he arrived in town, the market was far below the price set by El Rey's queen, and he did not dare to sell. Day by day the price dropped, and finally he sold in desperation. And then he was faced with the prospect of going back and telling his wife. His courage failed him; he simply could not face that; so he stayed on in town week after week.

Each morning when the public rooms opened, there he was standing at the door, and soon he was seated and starting on a fresh bottle of rye. By noon the first bottle would be gone. Another bottle disappeared during the afternoon. Day in and day out, he polished off his two bottles, and often he would not feel like going to bed after dinner. Then he would down a third bottle during the evening.

He was never quite sober, yet he was never obviously

drunk. As the day wore on and the bottles emptied, he would become more and more silent and would hunch lower over the marble-topped table. His eyes would become glassy and as expressionless as those of a stuffed animal. By afternoon, if anyone spoke to him, El Rey would look up, trying to make some connection with the outer world, and then would laugh mirthlessly and turn again to his bottle. In two months he drank at least seventy-five quarts of whiskey, and he did not change perceptibly. He ate nothing but ham and eggs, one plate for breakfast, one plate for lunch, and one plate for dinner, seven days a week. He said that his stomach was weak and could not stand other food. And besides, wasn't he an American? He had almost forgotten English, and his oldest boy, who was in town with him, spoke only Patagonian Spanish.

This boy was a great favorite at the hotel. A serious little chap with straight black hair and a tawny skin, you would never take him for an American, although in his Spanish argot he used to insist that he was one.

And indeed he did come closer to being an American than did another who spoke English. This lean and shifty, sandy-mustached individual came lurching into the Colón about tea-time one afternoon. Tipping his greasy cap still further on one side of his head, he said that he had heard us talking American and that he was starving for the sight of fellow-countrymen.

Portland, Oregon, was, he said, his home town, and he had just arrived from there (afoot, presumably, as no ship was in the roadstead). More than suspicious, we asked him whether he had ever sailed into Portland Harbor— Oh, yes! Hundreds of times, and what a time they had beating around Hatteras on the way in!

"Why, you aren't even American!"

"Gor blimey if I 'in't!"

In a few minutes he was insisting just as vociferously that he was an Australian who had just come down from the Cordillera, giving the name of a company that was dissolved years ago and of a foreman who never worked for that company. Then he said that he was an estanciero up from Gallegos. Finally, having reached the maudlin stage, he said he was only a beach comber and never had had a home.

He was, in fact, one of the Homeless Men. El Rey, in a lucid moment, remembered having seen him wandering through the blue of the back country years ago, and Townsend, an estanciero from up north, said that the man had once drifted in on him and had tried to sell a pair of eye-glasses.

The official but self-appointed welcoming committee for all of English speech who reach Comodoro is Whiskey-proof. In this case, that is his real, fairly earned and proudly borne nickname. Being welcomed by Whiskey-proof is pleasant but expensive. He is, in one person, the Culbertson, the Lenz and the Four Horsemen of dice games, and only at very rare and unhappy intervals does he have to pay for any of the whiskey against which he is proof. From him we learned the noble game of bidou, and although the lessons were costly, they were so thorough that they later paid for themselves many times over.

For many years Whiskey-proof lived in Magallanes, which is the Chilean town at Punta Arenas, usually but incorrectly itself called Punta Arenas by those who have not been there. And Magallanes seems to be the world-capital of dice games. Bidou, itself, the king of these games, probably originated farther north (there is some mystery about that), but it has found hearty and able exponents in Magallanes, and there are innumerable other games which are not, I believe, known anywhere else. As to the best of these, always excepting bidou, I cannot go into detail because their necessary terminology, like so much of Scottie's life, is a little coarse.

Whiskey-proof is a store-keeper, or better, manager, for he seems to avoid the responsibility of any partnership. In Comodoro he was working for the Anónima, and in Magallanes, also he had worked in one of the stores.

There he once had the misfortune to have his store burned and was immediately taken off to jail on the theory that all fires are arson and that any suspect is guilty until proven innocent. Although technically incomunicado, he managed to send out word to his friends, who rallied around with gifts of clean bedclothes, food, and whiskey. The warder indignantly returned the whiskey, saying that prisoners could have all the beer they wanted but not spirits—one of his inmates had been allowed some spirits on a previous occasion and had proceeded to take the jail to pieces.

Forthwith Whiskey-proof's friends sent in bottle after bottle of beer. He happily opened one and took a great swig from the bottle, and then nearly fainted. Even he is not wholly proof against whiskey drunk as beer, and all his friends had conceived the bright idea of sending in whiskey in beer bottles. The warder was invited in and offered some of this liquid wolf-in-sheep's-clothing, which he gladly accepted and from which he only recovered after some fifteen minutes of choking and gasping. But he was a sportsman and admitted himself beaten, and he stayed in the cell with Whiskey-proof all night, drinking, then singing, and finally sleeping. At dawn, Whiskey-proof, still conscious and happy, took the keys from his now thoroughly sodden jailer, let himself out, and went home. The charges against him were never pressed.

That is not the only time that a uniform has been on the wrong side of the bars in a Patagonian jail. Not long ago there was a series of bold burglaries in Comodoro. After many vain efforts, the perpetrator was finally caught, tried, and sentenced to a few days in jail. He struck up quite a friendship with the police while incarcerated, and as soon as he was released he joined the force. From this peculiarly advantageous position, he continued his depredations against the citizens and their households, until finally one night he was caught red-handed, robbing a house while dressed in his police uniform. This last touch was a little too much to be overlooked entirely, and when we were in Comodoro he was back in a cell. The popular opinion is that he will probably be made a judge when he is released.

Whiskey-proof, as his name implies, is one of those peculiar souls on whom alcohol has hardly any visible effect. He would stay up all night, greeting visitors or wishing them godspeed, both occasions for somewhat hysterical merriment in the far south, and then work all next day quite successfully.

The possession of inhibitions of behavior sufficiently strong to resist practically any amount of drinking is a trait that seems exclusively Anglo-Saxon. The Patagonians of Latin or Indian extraction almost invariably become murderous in their cups, and so it is fortunate that they are, as a rule, less given to this vice than are the Nordics in their midst. Out in the country, the camp, sobriety is the rule, partly from economic necessity, but also and more from custom and temperament. In the city, the Latins are inclined to go on occasional sprees—as who is not under such conditions, the world over—but the steady drinking is done by the more northern races, which is to say here principally the British, numerically predominant, but also such North Americans, Dutch, Germans, and Scandinavians as may be about.

Drinking among such people in Patagonian towns is Gargantuan, fabulous, heroic. There are exceptions, and any of our friends who may read these lines are to number themselves as such, but the rule is alcoholic almost beyond belief. El Rey with his two or three quarts of whiskey a day is no more than a striking example of a fixed tradition and custom.

Even wine, the mild table beverage from Mendoza, is often drunk solely for its Nepenthean properties, and as if the chief purpose of life were forgetfulness, which indeed it may be after a few years of Patagonia. At convivial gatherings, wine is frequently drunk "al vacío"—"until empty." The glass is held for filling between the thumb and little finger, one on the rim and the other on the base, in such a way that it can hardly be successfully set down again until empty. At the cry of "Al vacío!" the goblets are drained at a gulp. They are immediately refilled, and so on until forgetfulness comes. Whiskey, the usual beverage, goes more slowly but about as copiously.

The "hair of the dog that bit you" is a very necessary adjunct to such a life, if work is to be attempted mornings, and this has become quite systematized. In Deseado (a smaller but much older and more typical town than Comodoro), the good, or less bad, hotels have each a waiter whose duty it is to get the guests out of bed in the morning. He goes to each room with a stiff gin-tonic, a large tumbler about half gin and the other half a prepared mixture of soda, fruit juices, and quinine, with a swimming slice of lemon. This is drunk in bed, and without it few of the old-timers will, or indeed can, get up.

But even this dipsomaniac atmosphere is, we were assured, only a pallid echo from the Good Old Days, the days when wool growing was a gainful occupation. Then, when wool was sold each estanciero had so much cash that it seemed to him quite impossible to spend it all. Champagne was drunk al vacío day and night. If it did not go fast enough that way, they would squirt it at passing dogs, wash their hair and their feet in it, or simply pour it on the floor to see it fizz.

To illustrate those now legendary times, an estanciero in Santa Cruz Territory told me of an occasion when he and two friends were in Deseado on their annual party. On the last evening they pledged themselves to dry up the largest bar in town, and set to it manfully, at first in the ordinary way. As they began to realize the magnitude of the task, they resorted to quicker methods. They set up bottles against the wall and shot at them. When they could no longer hit these marks, bottles were broken against the mirror and other fittings, until finally the stock was all drunk or broken and the furniture was reduced to kindling and fragments.

Next morning they left before dawn, and at crossroads they parted and each of the three friends went on toward his own estancia and another lonely year. My informant was soon overtaken by a hard-riding messenger from the bar's owner. He was told that he and his friends had done five thousand pesos of damage the night before, and would he pay it all or should they send after the other two also? He still felt wealthy and expansive, and he gave the messenger the full amount. In the meantime each of the other two had also been overtaken by other messengers, and each had paid the full amount asked for the damage done.

Next year they met again in Deseado, and they happened to compare stories. They decided that their barkeep owed them two more bars, so they moved in on his rehabilitated establishment and tore it to pieces even more thoroughly than the last time. The next day, with revolvers to cap their arguments, they persuaded the man to clean out his place and to fix it up once more, carefully checking the supplies that he brought in to see that he did not stint. Then they proceeded to wreck the place again, with real barbaric fervor, so that for a third time the anguished proprietor had to see his place suffer a triple human cyclone, and this time so thoroughly that it was practically annihilated. He had long since spent the money collected the year before, and now he was quite ruined and soon disappeared from Deseado.

"Taking to drink" is very much a matter of course in that bleak land. There are few who do not have some strong and tragic reason or excuse for seeking this desperate release.

One of the most heroic drinkers of my acquaintance there had come out many years before, with the usual vision of quick wealth and a triumphant and early home-coming. The wealth was delayed, as always, and as he saw his return eternally postponed he determined that he would at least have a partner to share the silence and the wind. He went back to Europe, married a young and attractive girl, and brought her to Patagonia. What arguments he used or what picture he painted to bring her here, I do not know. She surely cannot have envisioned the reality. She had lived in luxury in a great city. Now she lived in a hut of corrugated iron, wore clothes and ate food forever permeated with sand, had sheep for society, and forever and ever the ceaseless and unbearable wind for music. She escaped, penniless, to town, and was taken away by a traveling salesman, the first man she happened to encounter who was leaving Patagonia.

The deserted husband finally scraped together enough money to return to Europe, where he divorced the truant and in some wonderful way managed to marry again. Again he returned to Patagonia with a young wife. This one was made of stronger stuff, or was, perhaps, less enterprising. In any event, she hung on for a longer time while the desolation and the wind gradually destroyed her nerves. She stayed so long that her husband felt secure at last and began to feel happy with his otherwise dismal lot. Then, and ironically on a Christmas Eve, he came home and found her hanging dead and swinging in the wind.

Stark tragedy is often abroad, down there, and yet it is sometimes hardly more grim than the mere eventlessness and emptiness of life and the slow but relentless horror of hope forever deferred and finally dying.

In the last analysis, "How not to behave in Patagonia" is not to go there as an alien intending to wrest fortune from the bitter land. The poor puesteros who accept their fate, who live calmly and whose hopes are not over the horizon, are sometimes happy. The only men who gain spiritual victories there, are those who stolidly accept their kinship with the land and who fight it and, in an inarticulate way, love it as a savage equal and worthy adversary. The tragic figures are those who came to Patagonia with scorn and superiority and who stay in defeated desperation.

How to Play Bidou

The game of bidou is so essentially part of the Patagonian scene, that its omission would be unforgivable in any true portrait. It is, furthermore, fascinating and worthy of wider knowledge. Now its habitat is curiously restricted: it is played only by British and Americans in the coast cities and towns of southeastern South America, from Rio to Magallanes. No one seems to know where it came from or when. It is not very old—perhaps ten or fifteen years. Its distant kinship with poker and some other typically American games suggests North American origin, but as far as I know it is unknown in the United States. In spite of being distinctively South American it is not Latin, nor do the Latins care for it, preferring as a rule games of pure chance.

Any number can play, and the equipment is simple: one dice cup and three dice for each player and a number of counters, usually six per player when the game begins. And the last man to have any loses the game.

The highest hand is 1—1—2, which is called bidou, and the second highest is 1—2—2, bidet. Then comes three of a kind, with 6—6—6 high, down to 1—1—1, and then straights, from 6—5—4 to 3—2—1. Below that the values are numerical, and pairs have no value as such; for instance, 4—3—1 beats 3—3—2.

All players throw at the same time and each looks at his hand without exposing it. If anyone wishes to hold his first throw, he then stops throwing and announces the fact loudly enough to be heard by all. This does not obligate the other players to hold the same throw. They may stop on the second throw, announcing that they are holding it, or may go on to the third and last. All three dice must be thrown each time.

One player is captain, chosen for the first round by throwing once and exposing the dice, the player with the hand of highest numerical value being the captain. Thereafter the captain is the player who won the last hand, or, in the case of his going out or of all passing around, the nearest eligible player on his left. The captain's duties are to call the throws and to start the betting. When he picks up his first and second throws, he must say so, and no one is obligated to announce that he is stopping or to pick up his hand until after the captain has said what he is doing, a fact which makes the captaincy distinctly disadvantageous.

If all stop on the same throw, the captain leads the betting, and otherwise the player who stopped on the earliest throw and is closest to the captain on his left. Once a bet has been made, the subsequent players, in rotation to the left, may withdraw, picking up their hands without penalty, or may see the bet, putting up an equal number of counters, or may raise. After a raise, subsequent players who have not yet bet at all on that round may pick up without penalty, or see the original bet plus the raise, and may also raise back. After a raise the players who have already made or seen an original bet must either see the raise or withdraw, taking back the chips that they had put in and also an equal number from each of the other players that saw the original bet.

The captain may check, with the privilege of seeing a bet which is subsequently made, but once a bet has been made each player must see it or withdraw for that hand. If all check, the captain does not have a second opportunity to bet, and all hands must be picked up and a new round started.

When the betting is completed, the hand of the bettor or last raiser is exposed, and his opponent must either take the chips or himself expose a better hand. In every case the lowest hand takes all the counters, or in the case of a tie for lowest, they are divided evenly. If a bet is made but not seen by anyone, the bettor may discard one counter, regardless of the size of his bet. This counter is simply removed from the game and no longer belongs to any of the players, but if a player has only one chip, bets it, and is not seen, he cannot discard it and must take it back. If all the bettors in a given round have equally high hands, each discards one counter and takes his others back, and if in the case of a tie for loser the chips cannot be divided evenly, the surplus is discarded.

As a two-handed game, the idea is the same but the procedure is slightly different. Nine counters are used and at the beginning these are placed between the players. As long as any remain in this central neutral pile, the players bet from it, the loser acquiring the number of counters bet, until all nine counters have been played, when they are distributed between the two players and bet as in the larger game. In the two-handed game, no counters may be discarded. The player who does not wish to see a bet must take one counter from the other or from the center, even if it is the last counter.

Every port has its own variations on bidou. Many of

these relate to the position of bidet, sometimes next highest hand, as given above, sometimes played as intermediate in value between 1—1—1 and 6—5—4, and sometimes considered as the lowest possible hand, losing even to 3—1—1, which is the lowest hand as I gave them above. Another variation (iniquitous, I think) is to penalize refusal to see a raise still further by making the refuser withdraw his original bet and not merely the equivalent but one more from each of the other bettors. In some bidou circles it is even permitted to discard the last counter held by the bettor. There are other minor variations, but I prefer the Comodoro game, which I have outlined.

In all cases players who have got rid of all their counters are out of the game, and the others continue until there is a single loser. No matter how many were playing to begin with, the two-handed rules go into effect when only two players are left, except that each starts with the chips he then has and there are none in the middle.

Simple in outline, once the values are clearly in mind, bidou is a complex and highly skilled game in practice. The betting of concealed hands and the ability to raise brings in the possibility of bluffing and all the psychological features of poker. The great number of possible hands and their very different probabilities and values involve a great deal of judgment in betting. The varying number of players in the course of a large game also makes the probable value of a given hand quite different at different times in the game and demands still further skill and practice for proper evaluation. Experience and judgment are so important that in the long run chance plays little part, although it may, of course, determine the outcome of one or two games. The combination of skill with speed is unique, and I know of no more interesting and satisfactory pastime. The whole and perfect scene requires also something cool in a glass and the strains of a tango, but the game still has merit even when deprived of this delightful background.

CHAPTER VIII

THE SOUTHWEST POLE AND THE MOUNTAIN
OF SMOKE

AFTER a short Comodoro interlude, the journal continues:

Vuelta de Senguerr
Casa Hernández
Dec. 3.

I was awakened before dawn this morning by a native field mouse with labor pains. I expended all my temper and threw all my shoes attempting to preserve my austere quarters from the indignity of becoming a maternity ward for Patagonian rodents, but the creature persisted in the quixotic ambition of having her children born under paleontological auspices. She finally produced two lovely twins under my bed, but their lives and very nearly her own were immediately sacrificed to her mad aspirations.

We struck camp, loaded it all on the car, had one last meal of roast lamb on the shore of Lake Colhué-Huapí, and then were off, driving without incident here to the home of Justino's parents. We have spread our beds in the front room, beneath the bright blue guns of the painted navy.

A Nameless Inn at the
Southwest Pole.
Dec. 4.

Our plan was to have lunch at the Confluencia, where the Río Mayo joins the Río Senguerr, to find a nice rich bed of fossils this afternoon, and to pitch camp tonight. The plan was excellent. It didn't work.

From the Vuelta to the Confluencia is about twelve leagues

along the river, but the way is so sandy that a car cannot be taken by that route now. We added ten leagues to the distance to avoid getting stuck, and got very royally bogged down anyway. From the Vuelta, we went up onto the Valle Hermoso, then onto the Pampa del Quemado, somewhat higher, and finally onto the Pampa Alta, still higher—the usual arrangement of flat plains at different levels. Thence down Cañadón María to the valley of the Río Mayo, and down the latter to its mouth at the confluence. Within sight of the Confluencia we came to a bridge across a tiny trickle of water, a bridge constructed by taking the ends out of some barrels and laying them endwise in the water. If only we had had sense enough not to bother about this effete contraption but to hunt our own place to cross! But we tried it, and it let us down literally and hard, into the almost bottomless muck.

We drove in tent stakes on firmer land and rigged a double block and tackle, but our best efforts did not budge the car. Justino went off and came back with a surly devil on horseback, but even with a horse on the tackle it was no go, and the rider left us without a word. Finally everything had to be unloaded, no light undertaking, and carried by hand to higher ground. Then after much digging and one thing and another, with engine working hard and three men on the tackle, the lightened car waddled out. Everything was reloaded, and so after several hours we reached the inn which was within hailing distance the whole time, and we are now partaking of its hospitality.

This is surely the farthest south in hotel accommodations! A more remote spot could hardly be found, unless at the Pole, and this really should rate as a Pole itself, the Southwest Pole perhaps. Beside us runs the Río Mayo, a clear, cold stream, fresh from the now quite near Cordillera, but its waters might be vitriol so far as the vegetation seems to care. It flows through the same central Patagonian landscape, rocky slopes and pebble-strewn flats with a few scattered thorn bushes. East of us (for we have rounded its end and come up the other side) rises the Sierra San Bernardo, a forbidding and jumbled mass of bare green and red rocks. On the other sides the valley is hemmed in by the precipitous slopes of the high pampas.

The presence here of a self-styled boliche, or country inn and trading post, is inexplicable, for finding a traveler here would be like finding Livingstone in Africa. We are the first guests this year; quite possibly we are the only customers the boliche has ever had, and in spite of its aspirations it is in fact nothing more than a sheep puesto of the poorer sort, save for the presence of an unstocked and unpatronized bar and of a spare room. It is built of mud blocks, with the fashionable corrugated iron roof, and has a walled-in patio onto which the four rooms open. A few yards away is a shearing shed, also of mud, with two small brush corrals, and there is a well with fairly good water, thanks to the nearby river.

Provision for dinner at this hotel consists of a live sheep and permission to use the stove after the regular occupants are through with it. The sheep was rapidly converted into meat, and Baliña stewed us up a good mess as soon as he could get at the stove.

There are several people here now, as they are shearing, and all of these as well as Justino and Baliña are sleeping in the large family bedroom. Coley and I have the very dubious honor of occupying the guest room, a tiny little hole without any windows. The one bed is a wooden shelf with barrel staves for slats and raw sheepskins for springs, mattress, sheets, and blankets. Coley is sleeping there, and I have set up my own cot. The conveniences consist of one tin wash basin, its innumerable holes plugged with rags. The luxuries and decorations are composed of someone's montura (saddle, bridle, whip, etc.) piled in one corner, a string of blown ostrich and duck eggs hung on a nail, and some tattered colored clippings of the Argentine football teams of several years ago plastered on the walls.

The moon is nearly full again, the night clear but windy.

Pueblo del Ensanche
de la Colonia Sarmiento,
a place smaller than
its name
Dec. 5.

Early up, paid our bill (six pesos, about a dollar and a half), and so off for the fossil fields.

The first peril was crossing the Río Mayo, a goodish stream and equipped with neither bridge nor ferry at this normally quite trafficless point. Justino said that he had driven across it and was not afraid. But didn't he get stuck? Oh, yes, he lost a car here once, but it did no good to be afraid! Wading out we found that the water is deep enough to submerge the engine of the car completely, the bottom sandy, the current swift, and that the only practicable approaches on the two sides involved driving straight upstream in mid-channel for about fifty yards. We wrapped waterproofs about the engine as best we could, and with sinking hearts confided the fate of our further travels to Justino's experience. To our surprise, all went well. The car very feebly dragged itself out the other side, coughing and spluttering, and did not stop until it reached dry and firm ground.

It was discouraging, after this victory, to get very thoroughly stuck in mere sand, as we did later in the afternoon.

The whole day was spent coasting along the Senguerr, looking for fossil beds which an earlier explorer had reported here, but we did not find them (and later learned that he hadn't either). The Senguerr is a permanent stream, being fed by the snows and freshets of the nearby Cordillera, and its valley is an oasis in the midst of the desert. It flows between forbidding and barren cliffs, but the flat valley bottom along which it meanders is green and fertile. Although very narrow, averaging perhaps a kilometer in width, this strip of verdure is slowly being occupied by a primitive settlement, poor but vastly better off than the pobladores of the pampas.

To the northward, upstream, the valley eventually widens out, and here there is a tiny town supported by the meager band of verdure along the river, an overflow from Colonia Sarmiento on the other side of the Sierra. There are two stores, an inn, a school, and a house or two. We are stopping at the inn, a fonda rather than a mere boliche, an elegant and pretentious place with a row of six guest rooms in a separate structure opening onto the patio. I somewhat underestimated the luxury of the place and asked the proprietor if there was some shelter where we could set up our beds. He was much upset and said that he deeply regretted that all his rooms already had beds in them, but perhaps they could clear one for us! This was too much for Justino and Baliña and they have spread their blankets out in the patio among the pigs and chickens, but Coley and I are actually to be ensconced between sheets, and fairly clean ones at that.

Not bad for a town which literally is not on the map! Of course the buildings are mud tempered with manure, the floors are native clay, and—well, there are certain other drawbacks that I may charitably omit, but it is comparative luxury. This is fortunate, for spirits were at a fairly low ebb. The expected fossil field eludes us, and also everyone in the party has some minor physical troubles which pull down the morale.

Colonia Sarmiento
Dec. 6.

There is considerable traffic between the Pueblo del Ensanche and so on and Colonia Sarmiento, proper, so that the track is well marked and has had some work done on it in places. There is a ferry across the Senguerr, which would otherwise be quite impassable for cars or carts, a platform of planks laid over three pointed scows. Each end is strung to a pulley running on a cable across the river, and by adjusting the angle the current is made to hit the scows in such a way as to impart a lateral motion in either direction as desired. Our outfit weighs about three tons and the ferry dipped perilously under it, but made the trip all right. Baliña says that there was a bridge here once, but it was destroyed or dismantled, and the ferry-keeper assures us that there will never be another. The reason, I suspect, is the five pesos per passage.

After a long detour, to check up on the geological formations on this side of the river, we got back onto the regular track, which goes down a long descent in the mountains and then through the Puerta del Diablo—Devil's Gate, unoriginal appellation, but here peculiarly fitting. This mountain mass, north of the Sierra San Bernardo, is unusually diabolical. The word "mountains" conjures up memories of sweet streams, beautiful forests, sweeping views across snowy peaks. These are mountains too, I suppose, but they have no trees or water, nothing gentle and nothing majestic. There is no repose. Everything is contorted and tortured. It is a great, choppy sea of steep bare slopes, of tilted rocks and untidy piles of jagged boulders. Even the rocks seem unnatural, for many of them are bright green, a most unusual color for rocks. The green of trees and grass is natural, healthy, and comforting, but to see the same hues in the barren stone of this grassless and treeless country is oppressive. It gives one a sense of foreboding, a feeling that the world is not friendly and not quite sane.

These vivid slopes, with red and yellow also splashed across the green, are capped by thick frozen flows of somber and repellent black lava. The Puerta del Diablo is a narrow pass between two high mesetas capped with this lava.

The track, extremely rough and in places somewhat dangerous, goes over a pass in the northern end of the Sierra San Bernardo and then winds down a valley into the Cuenca de Sarmiento. Near the pass, at a boliche enticingly named "Las Pulgas," "The Fleas," a horse was tied up while its master drank or gossiped inside, and apparently automobile traffic is not heavy here, for it almost went crazy at the sight of us, pulled up the thorn-bush to which it had been tied, and ran ahead of us for miles. It outran us finally and was seen no more. Part of the montura fell off, and we left it at the second and last boliche on this road, "Manantial Grande,"

"Big Spring," which by the way, is built of stone, a material abundant here, but seldom used.

As we emerged from the mountains, I suddenly identified the element that had made them so indefinably sinister: it is their lifelessness. They seem dead, elaborately and painfully dead, and as if they had nothing to do with life. They are like the moon. Even in the midst of them, it is hard to believe that places can exist so nearly devoid of such life as we elsewhere take for granted.

Here in the Basin, life is teeming. It seemed like a vision of the primordial earth, before man ravaged it, and as we like to imagine it (wrongly, say the grim mountains behind us), swarming with fearless beasts and birds. The water of Lake Musters was littered with wild fowl, ducks of four or five kinds, abutardas, herons, and enormous black-necked swans, in size rather like sea-going ostriches, as Coley remarked. Being men, we shot two ostriches, a martineta, and two ducks on the way, and so were well supplied with meat and a prospective change from mutton.

Now we are at the Rojo Hotel. Having maté in the kitchen, we met the new cook. She is a great, red-faced, raw-boned female, and we were so rude as to make some slight remark in English about her appearance. To our great embarrassment she proved able to understand English, but she forgave us and launched into the story of her life, happy to have an audience. She is so accustomed to Spanish that her English is halting and sometimes hardly comprehensible, but it is her native language. She was born in the Falkland Islands, and took one short trip to England, but most of her life has been passed in Tierra del Fuego and Patagonia. She must have seen, perhaps done, many interesting things, but if so, they are permanently stored away, for she is too nearly inarticulate to communicate them. She chiefly wanted to know whether we knew a distant cousin of hers whose name she thought was Patterson or Patrick or something like that and who once lived on Hartford Street in an American city the name of which she had forgotten. She looked askance at us and, I am sure, decided that we were not really Americans at all when we confessed our ignorance of her relative. It is hard to imagine how people like that picture things in their own thoughts. Probably her idea of North America is of a place very much like the Falklands.

Our new camp is a few leagues north of the lake basin, in a natural amphitheater surrounded on three sides by cliffs in which fossil-bearing rocks are exposed. The exact spot is called Pajarito, Little Bird, and even is to be seen on one of the crude maps of this region. What an erroneous impression maps can give, with their multiplicity of names suggesting a dense population! In this case, there is at Pajarito one long-abandoned dug-out hut, even more primitive than the usual puesto. There is a spring, very dirty but perhaps it will clear, and in the meantime we have drinking water in a tank while this will serve for other things.

To the west rises another mountain mass, very black and forbidding across the wide valley, the Sierra de Castillo, having, of course, nothing to do with the here distant Pampa de Castillo. Our amphitheater is cut into the edge of an isolated peak, the "Cerro del Humo," "Mountain of Smoke." The name is poetic, but its origin, now lost, was probably very commonplace. Perhaps when the country was first settled someone saw a cloud of dust rising from it. There is "Laguna de los Palacios," north of here, which owes its poetic name, "Lake of the Palaces," to the prosaic fact that a family named Palacio once lived near there.

Nomenclature is still very confused here. In most cases there have been no official designations, and those that have been made, on government maps and the like, not infrequently are wholly different from local usage so that an actual inhabitant may not know the official name of the place where he lives. The nomenclature is also very inadequate, and physical features that would be world-renowned if they were in Europe may not even possess names.

The growth of a system of geographic names is interestingly visible here, where almost none of the names extend back farther than a generation. Little imagination is shown, and practically all of the names fall into four categories. A few, surprisingly few, are of Indian origin, such as Colhué-Huapí and Senguerr. Many more, like Cerro Blanco, Cerro Negro, Pampa Pelada, were applied by the European settlers and simply describe the things they name. This is a very confusing class, for the same names naturally are used over and over again for different places. I know of three different rivers called Río Chico, four peaks called Cerro Blanco, and so it goes. A third and also very common sort of name is derived from the surnames of families living in the region, families which are thus lifted to an odd sort of fame and whose names will linger on here like ghosts long after the men who bore them have been forgotten. Such are Laguna de los Palacios, Paso Niemann, Cerro Purichelli, and innumerable others. The last distinct class of names includes those that are commemorative, in most cases given officially or semi-officially and often in conflict with the more natural names used by the inhabitants. They seldom have much reason or logic in their specific applications. Some are holy names, as the Pampa María Santísima, others, like Comodoro Rivadavia, are names famous in Argentine history, and a few like Lago Musters or Cañadón Tournouër somewhat more aptly commemorate Patagonian explorers.

The passion for naming things is an odd human trait. It is strange that men always feel so much more at ease when they have put appellations on the things around them and that a wild, new region almost seems familiar and subdued once enough names have been used on it, even though in fact it is not changed in the slightest. Or, on second thought, it is perhaps not really strange. The urge to name must be as old as the human race, as old as speech which is one of the really fundamental characteristics by which we rise above the brutes, and thus a basic and essential part of the human spirit or soul. The naming fallacy is common enough even in science. Many a scientist claims to have explained some phenomenon, when in truth all he has done is to give it a name.

And what a lot of things and places there are to name, here in Patagonia, even now! After traversing this much of the country, it gives some impression of monotony, but this impression must be due to the barrenness of the land, for on analysis it is clearly false. The form of the surface of the region is quite remarkably varied, and the landscape is put together very intricately from a great many different types of single features.

The most striking of the elements in the anatomy of the Patagonian landscape are the elevated flat surfaces, striking because in the more densely populated parts of the earth such features are usually rare. The surfaces have no drainage, and are usually pebble-covered, and the edges are usually abrupt slopes leading to a higher or a lower level. Although all of these are similar in appearance they have three separate origins: a few, relatively narrow and less obvious, are terraces of sand or gravel piled up by rivers or the sea; some, rather more common but also small as a rule, are benches or mesetas held up by lava flows or other beds of hard rock, the fact that they are horizontal being more or less accidental; the great majority, however, including all of the large ones, are terraces or old uplifted plains originally cut into or planed down across the substratum by streams or by the sea. The pampas, so very typical of central Patagonia, have this last origin.

Then there are the mountain ranges composed of tilted and folded beds of rocks. In this region these usually run north and south and each is rather isolated, small as mountain ranges go and not very high, but extremely rugged. They do not owe their elevation and character directly to the folding of the rocks that compose them, but to erosion which washes down and away the softer rocks and leaves the hard beds jutting up. Folding does not always mean mountains, for sometimes, as in the western part of the Valle Hermoso, erosion has been so active that it has cut down even the hard layers and left a flat surface. The Sierra San Bernardo is typical of these folded mountain ranges.

There are numerous isolated peaks, the origin and character of which are very different from those of the true sierras. Some of these, like Pico Salamanca, are merely parts of a high plain, in this case the Pampa Castillo, which the incessant erosive forces of wind and rain have isolated from the parent mass. In this way is brought about the Patagonian anomaly that the peaks may be lower than the surrounding plains, as Pico Salamanca is now lower than the Pampa Castillo of which it was once a part. Wholly different in origin are the peaks, like Cerro Negro, Cerro Tortuga, Pico Oneto, and countless others, that have a core or a capping of hard lava. This lava resists erosion, and so as the general level of the country is worn down, these peaks are left standing up. They may occur anywhere, on the pampas, in the basins, even in the midst of the folded sierras, like Cerro San Bernardo which is an igneous peak in the midst of the sierra of the same name.

Those are the important positive elements, the ones that stick up. The principal negative elements, the ones that sink in, are valleys and basins. The valleys of central Patagonia are like those of almost any semi-arid land. In this region, outside of the Andean Cordillera, there are only two rivers, the Río Senguerr (with a few tributaries from the Cordillera, such as the Mayo) and the Río Chico, and these are really only one river, interrupted by the lakes in the Cuenca de Sarmiento. Both these rivers meander along narrow flood plains between steep valley walls, the Río Senguerr always with some water but the Río Chico dry in the fall and winter and occasionally dry for a whole year or more. The other valleys contain running water only immediately after a local storm. The cañadones, cut into the edges of the plateaus and other highlands, are frequently very large and deep and are

really grand scenic features, but they are short, the longest
not exceeding ten or fifteen miles. They correspond, at least
roughly, with the arroyos of the North American Southwest,
but the Spanish word "arroyo" is used differently in the
Argentine, being applied rather to mountain valleys with
permanent streams. Still smaller are the zanjones, which
we would call draws or coulées in our West.

The most peculiar depressions of central Patagonia are
the basins, hollows the bottoms of which are deeper than any
outlet. These are very numerous and highly characteristic,
varying in size from the great Cuenca de Sarmiento down
to small hollows a few yards across. When the water supply
is adequate these basins contain lakes, with fresh water when
there is an outlet, as in the case of Lakes Musters and
Colhué-Huapí in the Sarmiento Basin, and salty when evapora-
tion is rapid enough and the water supply small enough to
keep the lakes from overflowing. Other basins, particularly
the smaller ones, contain water only periodically and often
have salt or alkaline deposits. Most of the very small and
shallow depressions appear to have been scooped out by the
wind, but the great basins are due to movements of the crust of
the earth, the deep-seated, slow but long continued slipping
of great blocks of rock against each other. Thus the Cuenca
de Sarmiento seems to be a large area where the crust of the
earth has sunk, leaving a rim of higher ground around the
large hollow.

From this point of view, the barrenness of the region adds
to its interest. It exposes the mechanics of the scenery and
gives an opportunity for vivid insight into the operation of the
forces that mould the surface of the earth. Here before our
eyes a world is being created. Each elevation and each depression has a meaning, and all intricately related and bound
together to make up the whole of the very characteristic
landscape. To the esthetic pleasure of viewing scenery is
added a deep intellectual pleasure when it is considered not
merely as a combination of meaningless forms, but as the
inevitable outcome of a sequence of events through millions

of years and the result of the combination and long action of natural forces still visible.

Dec. 10.

This camp promises to be only moderately productive. In our three days here, we have found six or seven first-class specimens and a number of isolated teeth and odds and ends. Justino has found the lower jaw of a *Pyrotherium*, which will be a very welcome addition to the Museum collection and which in itself makes the camp a success, but on the whole the exposures are limited and not very rich in fossils.

There are no human inhabitants for many miles around, and wild life is very abundant and unafraid. At camp there is a perfect plague of cuyses, wild guinea-pigs, very like the tame ones in size and build but with short grey fur. On the trail to the place where I am now working there is a village of tucutucus, small rodents rather like pocket gophers. They derive their staccato name from the odd sound that they make. Most of their time is spent underground, but at the village this morning there were two babies out sunning themselves, appealing little creatures, soft and furry, almost earless, with long tails, and bodies hardly an inch long. Like all the creatures that live here in the big, clean outdoors, however, they are already extremely afflicted with parasites.

About five years ago Baliña wandered past this place, hunting ostrich eggs, and about two miles from our present camp he saw a fossil bone. After all this time he remembered not only the fact, neither important nor interesting to him, but also the exact spot, a tiny dot on the large face of Patagonia. He described it so exactly to Justino that the lad went right to the place and picked up the very bone that Baliña had seen. That is a striking example of the way memory takes the place of written records with such men. I and others like me have become so accustomed to relying on writing for preserving information that such a feat of pure memory would be quite impossible to us.

Justino awoke me this morning by shouting that the wind

was blowing, which is like telling a sailor that the sea happens to be salty today. The wind has been blowing hard every day for weeks, although occasionally it moderates at night. The fact that the air is in rapid motion becomes almost as elemental as the fact that there is air. It is a condition of life here so universal that every action that is aided or impeded by the wind becomes directional by second nature.

While Coley quarried near camp, the lad and I went off to examine some interesting cliffs northwest of here. There is an eagle nest there, and by climbing down the vertical rock face we managed to get right to it and to photograph the inmates, three babies. This is an unusual number, as all the other nests that I have seen had only two eggs or young. They are ugly little brats with large heads already completely aquiline, bodies covered with white fuzz in which black quills are just beginning to show. The housekeeping is very untidy and the nest was foul with the odoriferous remains of many meals. Today's tidbit was an armadillo, a pichi, which we saw the old man kill. Hares and martinetas also have been figuring largely in the babies' diet. These big eagles have a wing spread of about six feet and are capable of killing lambs. The prey is usually killed by a single strike with the eagle's sharp talons, and if this is not enough, at least in the case of smaller animals, the victim is dropped from a height and struck again immediately after it hits the ground.

The fledglings, about the size of spring chickens, were not at all disturbed at us, merely opening sleepy eyes when we handled them and then going right off to sleep again, too full of pichi to move. The old couple, however, were raving mad, soaring above us screeching at the tops of their voices, but they did not attack us and kept out of shooting range, to the disgust of Justino who wanted a skin.

Coming back, we saw two chulengos. Justino squeaked with joy like a schoolgirl and dashed after them with the shotgun, the only weapon we had with us. After a long chase, they escaped and the lad came back angry and crestfallen. Then he thought to look at the gun and found that it had not been loaded anyway!

A strange moment yesterday—I had gone out, alone, on a little mound of clay, splotched with red. A great billow of streaked black rolled ponderously up over the Sierra Castillo and the jagged lava slope of the peak spread blackly with menace. The treeless valley flickered and grew somber, while the striped walls of the badlands closed in silently. Even the wind paused for a minute and the world became deathly still and unbearably oppressive. A strangely silent guanaco on a nearby peak looked like a black human skeleton against the sky, looking down with hate on a spark of life in the midst of eternal desolation. I was alone in an uninhabited, sunless world, heavy with death, infertile as a stony tomb, animated by inactive menace, and I could only wait in desperation for the towering, quivering walls to tumble and to crush me.

Dec. 14.

After morning maté and a light snack of two or three mutton steaks apiece and fixings, Coley went back to his quarries and Justino and I warmed up the car and hit out for the high places. We went down to the north end of Lago Musters, then up Cañadón Matasiete, across the crest of the Sierra Castillo, and down into the Bajo de Buen Pasto. Encounters were few but interesting.

First came five chatas, loaded with baled wool on its way from the Cordillera to the coast. Each chata is drawn by six to twelve horses, spread out fanlike and in apparent disorder in front of the cart, each pulling on a separate line. The trip takes them three or four weeks, as they only average seven or eight leagues a day, traveling from five o'clock to eleven in the morning and from four o'clock to eight in the afternoon. The long noon stop is the usual siesta time, extended to give the horses a chance to graze, if they can find anything edible.

The second encounter was with a battered Ford truck to

the front of which was attached an enormous sign reading "Correo Nacional de Colonia Sarmiento a Esquiel"—the mail on its way to Esquiel, a distant and tiny settlement in the foothills of the Andes. Alongside the track there is also a telegraph wire that is supposed to maintain communication with several of these very remote hamlets, but the wire is broken and the whole line down in several places.

Our last meeting was on the very top of the sierra, with a great wind blowing straight from the Cordillera. A man passed riding a horse at a fast lope right into the face of the icy gale. He was wearing a poncho which flew straight back in the breeze, and his face was covered with a black sash, in which only a tiny slit for the eyes could be seen. He passed without so much as turning his head, like a phantom horseman condemned forever to gallop over that cold and bleak waste. The horse shied violently at our car, but the rider kept his seat without seeming to move a muscle.

Having seen the geological exposures that occasioned the trip, we turned back and reached camp well before dark. Justino shot a guanaco and started skinning it. I sat on the bank of the zanjón thinking philosophically about here today and gone tomorrow, what bloodthirsty brutes we are, and so on, but then helped with the skinning and became so interested in it that I looked around, unsuccessfully, for another guanaco to kill. Philosophy requires inaction.

Colonia Sarmiento
Dec. 15.

We struck camp this morning, loaded, and drove in to Sarmiento. This was Justino's last day with us, a sad parting as we have become very much attached to the lad. For this country and this sort of work it would be hard to imagine a more agreeable companion or more useful aid. His number has been drawn, and he has to enter the army for a year of compulsory service. This annoys him very much, because he is now so wrapped up in fossil digging, but he takes it cheerfully, as he does everything that befalls him. I only hope that the army does not rob him of his spontaneity and naïveté.

He insisted to the last that he was working for us entirely for the love of nature and for the glory of science, that we are friends and that he cannot take pay from friends. I pointed out that friends are permitted to give each other gifts and after repeating many times that it was a free gift and not pay he finally accepted two hundred pesos, about sixty dollars, which he has earned many times over. We are paying the cook a hundred pesos a month, and Justino has been incomparably more valuable to us.

At the northern end of Lake Colhué-Huapí, on our way to Sarmiento, we passed a large flock of beautiful rose and white flamingos, an unexpected touch of tropical color.

> Government surveying camp
> near Casa Williams
> *Dec. 16.*

On our way out of Sarmiento this morning we got even more completely and more nastily stuck than usual, in black ooze tinctured with a long dead and very aromatic sheep. Another car came along, dashed right into the same mess, and also bogged down above the running boards. The owner, an estanciero who is attempting the long and almost impossible drive from Magallanes to Buenos Aires, went off and returned with a very blond German colonist in a Ford, and after a couple of hours' work we were all successfully extricated.

The rest of the road, on which we were only stuck once more and not badly, was the familiar trail around the southern end of Colhué-Huapí to its outlet into the Río Chico, which we reached at sunset. Here we were delighted to encounter a friend, the Danish topographer von Platen, who plied us with beer and Bols and gave us a feast, with tablecloth, napkins, and all the comforts of home—well, nearly all. These government camps are almost too luxurious; it takes a small army of men to transport them, set them up, and care for them. Von Platen is surveying the valley of the Chico for the Y.P.F., the government oil corporation.

This place is called Williams, because a Boer of that name used to live just across the river. I once saw a map of central Patagonia which showed a large town, Puerto (Port) Williams here. The "Puerto" part was probably someone's misreading of "puesto." The "town" consists of one mud shack and the "port" is over a hundred miles from navigable water !

Cerro Salpú
Dec. 17.

Coming down the valley of the Río Chico, along a wagon track now quite abandoned, we managed to reach Peñalba without incident—Peñalba, by the way, is simply a place where there was once a house and is now nothing. There we came to a zanjón with soft sand, and on the theory that prevention is better than cure we turned to and built a road across it, digging out approaches and cutting brush and piling it on the sand. We built a really beautiful road, but we got stuck anyway and it took us four hours to travel the next hundred yards.

Thereafter the way was just a succession of zanjones, but they were fairly narrow and by rushing the car at them at top speed we managed to jump out and up the other side. We broke the front springs, but we got through. A few leagues beyond Peñalba there is a new concrete bridge across the river. Everyone is justly proud of this, but it has the slight drawback that there are no roads leading to it. It just sits there in the wilderness in solitary grandeur. We hunted up a nearby puesto and had maté with a deaf old man, a half-witted boy, and a young man who complained bitterly that they were all starving to death. The old man couldn't hear our questions, the half-wit couldn't understand them, and the young man couldn't answer them. He had never heard of Cañadón Vaca, which must, nevertheless, be quite nearby, did not know whether there were any trails on the other side of the river, did not know anything about the region, explaining that he had only been there a few years.

We finally crossed the bridge, which may quite well be the only time a car ever did, and struck out cross country down the river. Near Paso Noruego, an uninhabited spot where wagons can ford the river, we climbed out of the valley and drove across the pampa to the foot of Cerro Salpú, a small peak of intrusive volcanic rock. It was getting dark and we were out of water, so when we came to a trail that looked as if it went to a puesto, we followed it.

The puesto, where we now are, belongs to an old Boer with flowing whiskers. He was far from cordial, but finally very reluctantly gave us permission to spend the night. The house is of stone, flat blocks of lava from the black cerro above it, and in front there is a bedraggled but brave little flower garden. The assembled family, taking maté, consisted of an ample Dutch housewife, three assorted but uniformly unattractive daughters, and a small boy. Soon after, another daughter rode up on horseback, a horrible little creature with a harelip and deformed body. What lives these people must lead, especially the women! As a rule, Argentine women (rare out here in any case) are to be found only at puestos of much higher type than this.

Communication is very difficult, as the family speaks only the Boer variety of Dutch, with a very few of the most necessary Spanish words. We gave them two ducks that Coley had shot down in the river valley, thinking that they would be overjoyed at this delicacy, but they threw them away when they thought we were not looking. Dinner, which we shared, consisted of boiled mutton, cooked so long before that it was literally putrid, followed by mutton broth. For dessert, a bowl of hot mutton grease was brought in and bits of hard-tack were soaked in this and then avidly devoured.

I wanted to sleep out, but our host was shocked and insulted and insisted on our setting up our beds in the front room, where he and the small boy also slept. All five of the females retired into the tiny, windowless back room, which must resemble the black hole of Calcutta.

Base Camp 3
Dec. 19.

We slept neither long nor well. No sooner were we in bed than the numerous dogs set up an awful din. Paw arose in his long woolen underwear, lit a candle, and hung over the bottom half of the door, peering into the darkness. After a time a newcomer, a young man with a tender mustache, came stamping in, ate, talked awhile, and then crawled into bed with Paw and the boy.

At the crack of dawn all were up, and we had to be spry to be out of the way so that the women could appear. For breakfast we finished up the rotten mutton of last night.

Paw fortunately had heard of Cañadón Vaca and knew of a good spring on its rim to which he guided us, and here we have just finished pitching our camp.

We are down in a small zanjón, somewhat sheltered from the wind, on the south edge of the great cañadón. The spring, at the base of a small sandstone ridge, is good water and has a fair flow. There are many frogs and tadpoles in it, but they do no harm. Their presence in all these isolated, small waterholes in a vast dry country is mysterious, but I suppose must be due to the carrying of eggs from one spring to another on the feet of birds and mammals.

This place has a bad name with the neighbors, and Paw warned us to be prepared to explain ourselves. Earlier this year a band of Chileans wandered down from the Cordillera and camped here, hunting chulengos and anything else that came their way. Things did not go well for them, and they took to killing sheep for food. The owners of the sheep did not care for this idea, and after some futile protests they got together in force, armed themselves, and came and drove the Chileans out. The latter scattered over the country and it is feared that some are still around.

These unsavory previous residents lived in a clump of thorn

bushes some yards from our present camp, where they built rude brush shelters. It is plain that they were in bad straits, for they ate a good many guanacos, and no native will eat them unless he is starving.

The valley of the Chico, which we traversed yesterday, presents a striking contrast with that of the Senguerr. Efforts to colonize the Chico valley have failed, largely because there is very little silted land and the water supply is very uncertain. Now there are almost no inhabitants, at least as far as Paso Noruego, and the valley bottom is as sear and barren as are the pampas. Paw says that this whole region is hopeless and that those who have not already left stay only because they have no means of traveling.

But I think we are going to find many and good fossils here.

CHAPTER IX

CAÑADÓN VACA

OUR baby owl, picked up on our way to this camp, hates us with a deep and inclusive hatred. We drove past the residence of his parents, a hole in the ground, and they flew off and left him to his fate, which was to be adopted by us and brought along as a prospective pet. A pet, I fear, he never will be. He looks at us balefully with his bright yellow eyes, and backs away when we approach, spreading his wings and either clapping his beak rapidly with a sound like chattering teeth or hissing very loudly like escaping steam. He refuses to eat, and we have tried poking meat down his throat with a stick, but this is not very satisfactory. Baliña says that no one ever tames owls, but apparently this is because they are regarded as being very unlucky.

Ours has certainly not brought us bad luck, for yesterday we located far the richest fossil deposit we have yet found, over on the other side of the cañadón. It is a little hard to get at from camp, but with the car we can make it in half an hour or so. Today we worked out a satisfactory route, across the valley on an old wagon trail, past a puesto on the far side, up a steep narrow zanjón to the top of the Pampa Pelada, and then along it to whatever point on the cañadón rim may be above the place where we are working. The exposures are all in the very high and steep wall of the cañadón there.

This afternoon I was working and not paying much attention to anything else, when Coley suddenly appeared and

170

told me to look at the sky. Bearing down on us at great speed was an enormous cloud, covering half the sky and black as night. I had some specimens uncovered which could be damaged by rain, so hurried to cover them up, and then dashed up the barranca, as it was plain that no more could be done today. The storm broke before I got up the cliff, and I finally reached the car, drenched through and almost blinded by the rain drops, which hit like shot as they were driven along by the gale.

We decided to go back to camp and struck out across the pampa. It is perfectly flat and almost absolutely bare—hence its name, "Peeled Pampa"—even more so than most of the high pampas. The only landmark for navigating it is distant Pico Oneto, quite invisible in the storm. We drove a long time without getting anywhere and finally became worried, stopped the car, and walked some distance away from it to take a compass bearing beyond its magnetic influence. By the compass we were going in precisely the wrong direction, and could hardly bring ourselves to believe that the compass was right. I suppose that most people who get badly lost do so because they think their sense of direction is more reliable than a compass. Anyway we decided to trust our compass, against our instincts, laid out a course, and managed to follow it by periodically stopping for a new bearing.

This worked, and we came out quite accurately at the top of our zanjón, down which the car slid very sickeningly in the mud. We went on past camp to Paw's puesto. Paw's name, by the way, appears to be Niemenhuis or something like that. It had only rained a little there, although the storm could still be seen across the cañadón deluging our diggings and the Pampa Pelada. The young man who arrived so late during our night here, and who is the husband of one of the girls, deftly dispatched, skinned, and gutted a lamb for us, then got dressed up in elegant store clothes and dashed off to call on the nearest neighbors, ten miles away.

We had some maté, but conversation languished. The only one who attempts conversational Spanish is the harelipped girl, and the Spanish spoken in the Argentine by Boer girls with harelips might almost as well be Hottentot.

And so back to camp, after dark.

Dec. 22.

A new wire fence bars our way to work, so we had dug out one post, leaving it upright and sheep-proof, but loose so that we can get through. This morning, to our great surprise, the post was firmly fixed again, just one post in miles of fence and far out of sight of any habitation. As we were eating lunch, the owner of the fence turned up, one Duvenage, a Boer and apparently of the more intelligent sort as he speaks English and Spanish very well in addition to his native tongue.

He told us that he often rode the fence, and had hunted us down when he saw that someone had gone through because he has a neighbor whose lamb crop is oddly large each year, Duvenage's being correspondingly poor. He was friendly when we explained our business and our lack of interest in lambs except to buy them for food, but he warned us that we would do no good here. He absolutely insisted that he knew all about fossils and that there are none here. In the meantime, as he talked, we were each working on a fine skull.

He knows that exact spot well, for he spent a winter there once years ago. He was driving sheep cross country and had just reached here when it started to snow. He stopped in his tracks, threw up a brush shelter, and simply stayed here for four months without moving. In fact he never got very far, eventually building a puesto nearby.

Later in the afternoon I took a long ramble to check up on the geology and to estimate what there is to do here and how best to allocate our time. Few of the new exposures that I looked at were very good or rich, and I decided to give it up and return, but to be on the safe side I went on to the farthest exposure of all. There in fifteen minutes I located three skulls and half a dozen jaws, riches almost unbelievable in this generally nearly barren formation.

Dec. 24.

"Do you know who discovered America, Señor Doctor?" Baliña asked me this morning as I drank my first maté.

"Why, some say one and some say another, but most people think it was Christopher Columbus—Cristóbal Colón."

"Sí, señor. It was Colón. Did you read the book?"

"Yes, I did notice something about it in a book once."

"No, no, but a big book. The book of Colón discovering America. It has the whole story. You should read that, you are always reading and writing things."

"I believe I haven't seen that book. Have you read it?"

"My father read it, Señor Doctor, many years ago, and he told me about it. That was why I wanted to come to Buenos Aires as a boy. To see where Colón landed."

"But I always thought that he landed in the West Indies!"

"No, señor! Montevideo first, and then Buenos Aires. You should read the book of Colón discovering America, and then you would know these things. May the señor doctor pardon me, but his Grace must see how ignorant he is. I will tell you, and you need not be mistaken again.

"You see, this Colón was a poor boy from my father's province in Spain. The queen, Isabella that was the queen, she advertised for a brave man to discover America for her. So Colón answered, and he got the job."

[Some of the details of just how he came to get the job must, I am afraid, be omitted. The word for "egg" has a double meaning in Spanish.]

"They sailed and sailed. But days and days, señor! No land. But days and days more, señor! Still no land. The food was all gone save a few galletas no one would eat. The water, she all turned green.

" 'We starve!' cried the sailors.

"Wicked men, they decided to eat one of the company. So they drew straws, to see who would be eaten. Even Colón was made to draw a straw, and the lot fell to him. He was annoyed."

"I would have been annoyed, too!"

"But wait, señor. He was not eaten. No! A little cabin boy spoke up.

" 'We cannot eat Colón,' he said. 'Use your thick heads,' he said. 'Colón is the only man who knows where America is,' he said, 'and if we eat him we shall be lost.' He was right, señor."

"Indubitably. So whom did they eat?"

"They ate no one. Pucha! Imagine to yourself! While they were arguing Colón said, 'That's all right. We don't need to eat anyone, for there is land. *Monte video.*'

"Of course he should have said 'Monte veo,' but he was an ignorant man and could not speak good Spanish. So he said 'Monte video' when he meant 'I see a mountain,' and that is how Montevideo got its name.

"They all landed and they killed enough steers to provision the vessel."

"But there were no steers in America until they were brought here after Colón."

"Do not interrupt, please, Señor Doctor. I am nearing the grand end. They killed steers and provisioned the vessel, but the natives were hostile and there was much fighting and many sailors were killed. The natives of Montevideo are still like that. Those men did not like Uruguay.

" 'Never mind,' said Colón. 'There's a much better city even closer than this on the other side of the river.'

"They crossed the river, and as they came to the other shore the wind began to blow in every direction at once.

" 'Varios aires,' said Colón, meaning that the wind was blowing in all directions at once. The sailors misunderstood him because of his accent, something like yours, Señor Doctor, and they thought that he said 'Buenos aires.' That is how Buenos Aires got its name.

"Everyone liked that city very much, but after a while they got homesick and they went back to Spain. Now, Colón had a common sailor whose name was América, and América sneaked off to see the queen, and he told Isabella that Colón had not done any of the work. América said that he had done all the work. So the queen gave him all the money and she named the country after him. That is how América got its name.

"Colón went to Isabella, and he said, 'Well, I have discovered the Argentine for you and I have come to get the money!'

"But the queen put him in jail and poor Colón died there. And that is how América was really discovered, Señor Doctor. It is all written down in the book that my father read."

Feeling edified, we ate breakfast and then Coley and I hit out for Comodoro, for tomorrow is Christmas and I feel a holiday coming on. Although our new route into town crosses the Pampa Pelada, the Chico Valley, and the Pampa Castillo, it is somewhat better than most we have had and we made good time.

Christmas Eve in Patagonia: I should wax poetic, but the appropriate feelings are all sad, and the words will not come. We finally spent the evening at the Bar Germania drinking beer. The orchestra, two girls and three men, all Germans, was playing jazz on a parlor melodeon, accordion, two fiddles, and bass viol. We finally bribed them to play "Stille Nacht" and "Tannenbaum," but the spirit was lacking and we had finally and sadly to give the whole thing up as a very bad job and go to bed.

Hotel Colón
Comodoro Rivadavia
Christmas.

The hottest day so far this summer. If there were any trees, palms would look more natural than Christmas trees. But there are no trees. There is really no Christmas either. Picking this as a holiday was a mistake; we are all exiles here, and Christmas is simply a day that has the power to drive despair more deeply and bitterly into the exiled heart. We have had a good time, but rather hysterically and in a spirit nearly the opposite of that appropriate for the occasion.

In the afternoon we went out to Manantial Rosales, the Anglo-Persian oil camp, and there had tea with the manager, Smellie, a quiet and very pleasant Englishman. Thence back to the hotel to dress for dinner. Even the English here do not quite carry out the legend of dressing every evening in the wilds to keep up the old British spirit, perhaps even that spirit is not equal to Patagonia; but we did dress this evening, and rather silly it seemed, too.

We dined at Ocho, I at Watts' and Coley with the Hallets, and then celebrated at the Club Petrólea. Ocho, so called because it is eight kilometers from Comodoro, is officially the Companía Ferrocarrilera de Petróleo, an English company chiefly engaged in producing crude oil for the English-owned railways. Most of the officials, and many of the underlings, are British. Like most of their countrymen they try as far as possible to overlook the detail that they are in Patagonia and attempt to live and act as if this were just a slightly out of the way corner of England. There is something splendid in this insularity, this refusal to accept any standard but their own, but it sometimes works out badly, at least down here. After all, they are in Patagonia and the country influences everyone whether he pretends to ignore it or not. Especially if he pretends to ignore it, sometimes. The more pliant tree outrides the storm, and I think the more adaptable spirit has more chance to survive Patagonia.

There is the manager, a charming example rather than a horrible one, and yet the country is visibly corroding him beneath his British exterior. A young man still, he started with nothing and rose by sheer brilliance to his present really important and very well paid position. His life is a regular, and I should think a dismal, cycle of pulling himself together and going to pieces again. He begins to drink, first a little and then as weeks pass, a lot. Finally he boards a plane and goes off to Buenos Aires, where he disappears for a while. Then he comes back, looking haggard, nerves frayed, and no longer drinking. This "water-wagon period," as he himself calls it, may last for a few days or weeks, and then the cycle starts again. His doctor tells him that he does not have long to live.

At his job, this man drives himself mercilessly, and he is a terror to his subordinates. For instance, while we were there, there was occasion to get into a storehouse. The man in charge was desperately ill, in bed with a high fever. Anyone else could have got the key, opened the storehouse, and done the necessary business in five minutes, but the manager insisted that this must be done by the man to whom the duty had been assigned, and the poor devil was routed out of bed. He did as he was told, and then went back and collapsed in a delirium. After hours, the manager had nothing but consideration and amiability for this same man whom he so nearly and needlessly killed in the line of work. The manager is a staunch friend and a charming host.

Wherever two men of English speech are gathered together, they build a golf course, and wherever there are three, they establish a club. So of course Ocho has both. The course sprawls over barren hills, with greens which are such in name only, rolled sand and crude oil. The club is a comfortable brick building with a library-ballroom-lounge, a game room (devoted largely to ping-pong), and, inevitably, a bar.

It was at this club that the Christmas party assembled after dinners at various homes, and there we had dancing and drinking and rather forced Christmas-spiriting until an early hour. A flashlight group picture will be fit only for suppression. It was taken just after the English bank manager's trousers had playfully been removed, with his wife, the only Argentine present, ineffectually trying to prevent this indignity and looking deeply bewildered and very injured. Their slightly superior attitude toward marital fidelity makes the English and Americans seem desirable as husbands to Argentine girls, but once married these usually try to *acriollar* their husbands, to Argentinize them, and they do not understand and are inclined to resent the less formal and, on the whole, happier Anglo-Saxon temperament, and are usually intensely jealous of the husbands' contacts with their own race. The picture caught others, including (I am sorry to say) me, in almost equally compromising and hardly more dignified poses.

The party finally broke up, we stopped at Davies' for a rather supererogatory nightcap, and so to bed.

Camp 3
Dec. 29.

On our way back to camp we stopped at Solano and had lunch with Deckers, whom I had not seen since my first days in Comodoro. He proudly displayed a new daughter, María Elena. Señora Deckers is an Argentine, handsome and charming, but clearly a little resentful toward us. To her we represent the alien in her husband, which she cannot fathom and so would like to eradicate and forget. The bank manager is distinctly subdued by the same attitude in his wife, but not so Deckers, whose incessant high spirits and joie de vivre are quite irrepressible. So Señora Deckers gracefully fulfills her duties as hostess and then retires and leaves us to our strange foreign ways.

Later, on the way home, we stopped at Duvenage's and bought a lamb, four pesos. His puesto is just below the rim of the Pampa Pelada, on the Río Chico side. It is the usual shack of corrugated iron with an air of abject poverty. The one unusual feature is a small plot of coarse and sparse grass, which he has fenced off as a ram paddock. The first time I saw a clump of this grass I was so overcome at the sight of some vegetation that looked soft that I proceeded to flop down on it luxuriously, only to arise more rapidly with howls of anguish. It might have been cactus from the results; even the grass here is tough and sharp. This patch of grass is the indirect source of the name Cañadón Vaca, Cow Canyon. An early settler ("early" here meaning at most twenty-five years ago) thought that the grass presented an amazingly fortunate opportunity to keep a cow, and he imported one with great difficulty. It nibbled mournfully at this harsh pasturage for a few days, and then starved to death. Its bones still lie there, scattered and bleached, and it is immortalized in the name Cañadón Vaca, so extraordinary in a cowless country.

Duvenage's wife is of his own nationality, Boer, and the balance of power is very different from that in the households I have just been discussing. She is a thin, sallow, faded, silent creature, so used to being alone that she is almost like a furtive wild animal before strangers. At meals, she sets his food silently before her man and then disappears until he is through eating and has gone outside. There are two small children, happy but very unhealthy-looking and extremely dirty.

Although really intelligent, hard-working, and even progressive, witness his many miles of good fences, Duvenage has barely managed to wring the scantiest necessities from this stubborn soil. No one here succeeds in doing much more, but Duvenage's particularly low estate is also due in large part to the peculiar fatality that dogs some men so persistently that it must come from within them. For instance, a few years ago he had managed by desperate efforts to build quite a large and fine flock of sheep. When they came to dipping the sheep one summer, he and his partner, one Norval cr Norwal, each unknown to the other put the requisite amount of chemicals in the solution, and they ran the whole flock through before they noticed anything amiss. All the sheep died.

Although he is a Boer and his name is French, Duvenage pronounces the name as if it were Spanish, Doo-veh-na'-chay (the *ch* a guttural much as in Scotch or German). The most striking relic of his origin is a whip made of a single piece of rhinoceros hide, originally about sixteen feet long and now worn down to about eight.

Like so many puesteros, Duvenage made capital investments in slightly more affluent days which are now useless because he cannot afford to keep them going. He has a Ford truck, but he has not run it for several years because he cannot buy gasoline. We give him some in exchange for sheep and for his work when we need it, and he is happy as a child at the prospect of being able to truck his own wool. He also has a Mauser rifle, which he has not shot for years because he cannot buy cartridges, but we cannot conveniently supply him with those. He has three or four cartridges hidden away which he is saving for some vague necessity. Probably he will never fire them.

During the summer, most of his time is spent in chasing guanacos. He hates them passionately, because their food is the same as that of the sheep and each guanaco on his land means one less sheep for him. He has grazing rights to three leagues of land, about seventeen thousand acres, but this enormous area is really hardly enough to support one man, let alone a family, for it takes about fifty acres of such country as Duvenage has to keep one sheep alive. Of course he does not own this property, and if he moves away he cannot even sell his fences.

The fences serve well enough to keep the sheep in, but cannot keep the guanacos out. One day we saw one of these agile beasts standing by one of Duvenage's fences. We startled it, and without any run it easily jumped sideways over the fence and loped away. When they are running, guanacos simply take the fences in their stride, as if there were no barrier there. Ostriches, on the other hand, are completely stopped and often killed by wire fences. They can neither jump nor fly and they frequently run full tilt into the wires, breaking their legs or hanging themselves. The fences are festooned with their dismal carcasses, and if the land were all enclosed, the big birds would soon become extinct.

The guanacos cannot be fenced out, running them down with boleadoras would be a hopelessly long task, and the price of cartridges prohibits shooting them. The alternative, adopted by Duvenage and all his neighbors, is to run them with dogs and on horseback, and he spends a large part of his time doing this. One day he will chase hard from dawn to dark and run most of the guanacos over onto the next man's land. The next day the neighbor chases the guanacos back onto Duvenage's grazing property, and so on, endlessly. It is splendid exercise for the guanacos, who seem to enjoy it and to thrive on it, but otherwise it all seems a little pointless. Occasionally the dogs kill a guanaco, but only when they have luck and are hungry enough to make a supreme effort.

The hunting dogs, galgos, and the sheep dogs, ovejeros, belong to two different packs and are not allowed to have much to do with each other. The galgos are mongrels mixed, so to speak, on a base of hound. They are encouraged to feed themselves. They are severely punished if they go anywhere near the sheep. The sheep dogs are mixed on a base of collie, usually, although they vary even more in type than do the galgos. They are never allowed to hunt anything and are fed at home. They naturally vary in training and value, but most are extremely good and I imagine that the best could compete with sheep dogs anywhere.

It is a real delight to see the sheep dogs work. They are so skilled and business-like, so anxious and able to carry out the slightest command, and yet they have an air that they consider themselves as partners with their master, and not his servants. We went out with Duvenage when he picked out our lamb for us day before yesterday, and he used one dog. It kept close to heel till we were up on the pampa. Immediately we sighted a bunch of sheep, the dog lay down and looked at Duvenage for orders. As soon as he said to go round up the animals, the dog simply streaked away, belly close to the ground and tail streaming out behind. When it overlooked a sheep, in a hollow or behind a thorn-bush, Duvenage would whistle and point, and the dog would unhesitatingly pick up the straggler, urging it along by barking and nipping its heels. Soon all the sheep of this point ("punta," a division of a flock that keeps more or less together) were gathered into a compact circle. Duvenage then looked them over and pointed out the particular lamb that he wanted, and the dog separated out the lamb and brought it to us, or rather made it come, meanwhile keeping the rest of the point entirely in hand. The dog was having the time of his life, and plainly felt as superior to the sheep as if he had been human. The only thing he hated was having to let them go. He likes to see them tidily bunched together and under control, and he was very crestfallen at having to let them scatter again untidily, but of course he obeyed Duvenage's command. Duvenage talks to his sheep dogs in Afrikaans and to his galgos in Spanish.

Incidentally, we once had a lot of trouble catching a lamb and Coley finally shot it, with the owner's permission, of course. This plunged Baliña into the depths of despair. The idea of shooting meat! ("Meat," unless qualified, is always mutton here.) What is the world coming to? Who ever heard of such a thing? It isn't *costumbre!* We will all end in a madhouse! And so on, far into the night. The proper procedure is to cut the sheep's throat and let it bleed to death, a nasty business but apparently not very painful.

We are back at work now, after our Christmas spree, and having excellent results.

Dec. 30.

The usual long day at the barranca, without incident.

Baliña, who is getting decidedly moody and cannot be depended on for good humor as at first, was in unusual form today. He is beginning to alternate between spells when he will hardly speak to us and others when he is abnormally gay, almost feverish. This morning he regaled us with one of his long tales, *vide* the History of Colón Discovering America for the general style. In this one, as in many, he himself was the hero. To make a very long story short, it seems that there were some robbers here in Chubut and that they slaughtered a family (many gory details). This was considered rather uncalled for, so police were sent after them, but failed. Then an army detachment was sent, and they were defeated by the robbers in a pitched battle. Then Baliña took over and formed a civilian posse. (Here a long and circumstantial account of the chase, with name and description of each landmark passed.) The bandits were finally cornered at the foot of a cliff and Baliña gave the order to charge, but at the first volley his posse turned and ran. Nothing daunted, Baliña charged all alone, captured the bandits single-handed, and took them into town where they stood trial and were executed. He seems earnestly to believe this, and perhaps it has some basis of fact into which Baliña's vivid, probably disordered imagination has written his own part.

"That cowardly posse," says our historian, "was Argentine and I should have known what to expect."

Although he has lived here so long, he is still fiercely Spanish and can see no good in the Argentine. He is especially contemptuous of Argentine industries and manufactures. "Industria Argentina," he says, should read "Made of Cardboard." This has become a camp joke. If the meat is tough, we say, "Well, it is an Argentine sheep and is made of cardboard!" Pride of birth is a curious thing. Baliña barely remembers Spain and in speech, appearance, and habits he is as entirely Argentine as the most native gaucho criollazo, yet he is filled with scorn for those from whom he differs only in the accident of birthplace, and they are just as scornful of him and call him "Gallego." That means Galician, yet it is used as a term of opprobrium and is a fighting word as applied to Spaniards in general when, like Baliña, they happen not to be Galicians.

The old boy has also been having fun playing in the water. First he dug out the spring, a necessary job which uncovered all sorts of relics. The spring has been in use since very ancient Indian times, for an Indian bola and an arrowhead, both of crude and ancient type, were found in it, as well as the remains of ostrich eggs and bones of almost all the native animals. Having thus insured a flow of good drinking water, Baliña has proceeded to dig and dam along the overflow until he has a series of four small ponds. He solemnly assures us that the first is for us, for bathing, the second for flying birds, the third for ostriches, and the last for guanacos. We are thinking of putting up signs and levying fines on anyone patronizing the wrong puddle. The guanacos never make a mistake, but this may be because their pond is the one farthest from camp. They come down to drink every morning at sunrise as many as fifteen and twenty at a time.

Inspired by Baliña, we suggested a Keep Patagonia Tidy campaign. Motto: "This is your Patagonia, help keep it neat." Program:

1. To provide trash cans for picnics.
2. To spray the badlands with Flit and keep them nicely dusted off.
3. To delouse baby tucutucus.
4. To provide the guanacos with curry combs.
5. To give all ostriches doses of vermifuge.
6. To establish Good Housekeeping classes for eagles.
7. To keep papers from blowing away, out of consideration for the people in Africa.

There was some talk on an eighth point, to plant a tree for the edification of the many inhabitants who have never seen one, but this was abandoned as quixotic, in favor of the more practical plan of growing our tree in Brazil and posting photographs of it at intervals along the Patagonian trails.

We did actually see what looked like a picnic on our way in to Comodoro the other day, and we were, and still are, completely baffled. In a hot, open, windswept, dry spot in the midst of the pampa, several leagues from town, a number of wagons were standing and around them were clustered whole families in their Sunday clothes, apparently doing nothing.

When I was first planning to come down here an explorer who apparently knew his Patagonia told me that the climate was awful, but there was one blessing, there were no flies. When I see him again, I hope that there is no weapon handy. There are unbelievable millions of flies and they almost drive us crazy when we are working in hollows in the badlands where the wind does not strike with full force. There is very little choice between being sheltered and having flies and being exposed and having the wind, with the attendant misery of volcanic dust in our eyes all the time.

For that matter, there are not only flies but also ants,

spiders, ticks, fleas, lice, mosquitoes, scorpions, beetles and nameless biting crawlers and fliers. This morning after dressing I felt something crawling inside my trousers, and disrobing hastily pulled out a scorpion several inches long. These are very common, and they love to crawl into our beds and clothing. A small spider laid some eggs in the cook's neck, which has swollen and festered badly, and something, probably the same kind of spider, has bitten me very severely around the waist so that I itch like fire. There are also enormous hairy spiders, like tarantulas, which are probably poisonous, but no one has been bitten by one yet. The scorpions, fortunately, are rather sluggish and no one in camp has as yet been stung by them. They probably could be quite painful, but hardly dangerous to life. There are also very natty large centipedes. The only disagreeable thing that Patagonia seems to lack is snakes. There are said to be some, but so far we have not seen the slightest sign of any. In the same sort of country in North America we should see rattlesnakes every day. That is a curious thing, too: rattlesnakes are common in South America but here they are typical of the tropical jungles rather than of the deserts. Fer de lance and coral snakes are said to get down into the pampas region, but if so we must be understudies for Saint Patrick. We have not even seen any harmless snakes.

Jan. 2.

The new year has started badly, with heavy downpours. Steady rain makes our field work impossible, and the frequent showers since we came to this camp have hindered us a good deal. With the customary animosity of Patagonian weather, it usually waits until we are miles out, with several mudholes and watercourses between us and camp, and we have been stuck frequently, so frequently that I have not bothered to make a record. The last forty-eight hours we had continuous rain, all day and all night, so stayed in camp. I got my notes all written up and for the first time since starting work in Patagonia had some free time in camp. I spent it reading "La Corbata Celeste," a historical novel by the famous Argentine writer Hugo Wast. I found it very interesting and well written.

Today has been fine for a change, sun and what would be a wind elsewhere but is only a refreshing breeze here. Myriads of flies, of course. Our trail is very soft, but passable. One specimen had to be rebandaged, and another was slightly broken, but the rain had not done any serious damage.

In the afternoon Duvenage turned up on his way from a lion hunt. Lions, as the inhabitants call the pumas or mountain lions, are still fairly common here in spite of having long been hunted. Where we are working we have frequently found their tracks in the morning, and once, perhaps as a peace offering, the carcass of a sheep which a lion had killed. Duvenage and the Van Wyks, who are the puesteros along our trail in Cañadón Vaca, have been hunting for this family for a long time but hadn't found it. The rainy weather finally softened the ground so that the beasts left tracks and they were followed to their lair, a small cave in the rocks just below where we are working.

The hunters walled up the back entrance and then sent in the hunting dogs to drive the lions out; but a full-grown puma on its ground is a match for any number of dogs and these soon beat a rapid retreat, howling and licking their wounds. Finally flaming brush was thrown in, and as the cats came out two precious cartridges were spent on them. There were two lions, an unusually large old female, seven and a half feet from nose to tip of tail, and a male cub. The Van Wyk boys took the large skin, and Duvenage carried off the smaller, slung over his saddle. North American horses as a rule will not carry a puma skin and I was surprised that this horse did not seem to mind it, but Duvenage said that it is an unusually tractable horse and that his others would not have carried it.

Jan. 4.

We stopped at Van Wyk's this morning. Hitherto we have just sailed past, hoping that they were consumed with curiosity, which they were not, having long since learned all about us from Duvenage. The old man is a very portly, clean-shaven Boer who moves slowly if at all. He speaks a few halting words of English, but apparently understands practically nothing said to him in that tongue. His Spanish is even worse, but the children have a little Spanish and no English. There are three children there now, and some others are away. All of them have a sullen and stupid look which I cannot help associating, unfairly perhaps, with the Boer colonists here. There is one girl, probably about fourteen years old. She wears very short skirts which show off her ( )-shaped legs to advantage, hobbles about the puesto in ridiculous, tawdry high-heeled shoes, and uses a thick coating of crimson lipstick. In spite of all this she is so shy that she ran when we stopped, and peeked at us covertly from around corners. Amazing apparition!

I bought the big puma skin from one of the boys, on condition that he cure it properly.

Jan. 7.

It still continues to rain at frequent intervals, but mostly at night so that the work has been going forward steadily and with remarkably fine results. Finding this place was very fortunate, and our collection from here alone will be worth our whole effort.

It rained off and on all night, and we could not sleep. Steady rain is soothing, but with these passing showers we just drop off when it stops and we wake, and then we are just getting to sleep again when the drumming on the canvas over us starts once more and awakens us. This afternoon it rained hard in camp, but at the barranca we got only a few drops.

Coming home last night we caught a young *perdiz colorada*, so-called "red partridge" but actually a species of tinamou, and this morning I photographed it and then released it where we picked it up, whereupon it scuttled off happily. Almost every day we race against ostriches and guanacos, both being abundant in this region. So are skunks, pumas, armadillos, hares (both real and the Patagonian pseudo-hares), and dozens of other sorts of animals. In spite of the settling of the country, if a family every ten or fifteen miles can be called settlement, wild life is still extraordinarily abundant. To visitors like ourselves who are here only temporarily and who like animals, this takes much of the curse off the country and gives it unending interest.

CHAPTER X

CHILDREN OF PATAGONIA

PATAGONIA might have been created expressly for poor, misshapen, muddle-headed beasts.

"Here," I can imagine their god saying to them, "here I have reserved you a corner of the earth, all to yourselves. It's a little lacking in comfort, perhaps, and you may find that you have to scratch pretty hard for a living. You'll have to put up with that, and you'll find that it's really a disguised blessing. If I gave you a country that anyone else would have, then smarter beasts than you, and especially that prime animal, man, would soon come and take it from you.

"You haven't much sense, poor dears, and in a world where brains count you'll never get anywhere. But I'm giving you a barren, wind-torn, worthless land where you'll be let strictly alone and where you can enjoy yourselves in your own way."

The plan did not quite come off, for it did not reckon with the perversity of man's nature which makes him deliberately seek out such bleak and bitter places, but it is not a complete failure. To this day man remains an interloper in Patagonia. He cannot look about him there, as he can in more familiar places, and say, "I am the owner and the boss here. What I did not create, I have subdued and modified to my own ends."

We men speak, a little patronizingly, of our motherlands, of the soil that gave us birth, and as precocious children we proceed to make our poor parent toe the line and behave as we wish. No man, however, can justly claim Patagonia as his motherland. At the best it is a stern foster-mother, and it gives more orders than it ever receives. Even now, the true children of Patagonia are the muddle-headed beasts.

It often happens that favoritism is shown to just that

child that seems to outsiders the least handsome and the most stupid, and the favorite child of Patagonia is surely the guanaco. A guanaco looks like a small, humpless camel, which it is, and it also looks like a careless mixture of parts intended for other beasts and turned down as below standard, or like the result of a long period of miscegenation. It has a head something like that of a hornless deer, long ears like a mule, a neck that tries but fails to reach the giraffe standard, a scrawny, shapeless body, and gangling legs like those of a young colt. To top off the joke played by creation on this poor beast, it has a stubby little brush of a tail, only a few inches long, which it carries crooked, the base vertical and the tip curved back and down, so that it looks very much like the handle of a jug.

Its back and sides are woolly and in the newborn are soft and silky, but within a very short time, even a few days, the wool becomes unpleasantly matted and in the old guanacos it is patchy, coarse, and filthy. The other parts have short, straight hair. The under parts are white, and this color extends up in streaks near the legs, while more exposed parts of the body vary from a rather dark russet to pale yellow. The forehead is usually grey. The animals are small, in comparison with most of the camel tribe, and slender in proportion. Even with their long necks and legs, the adults usually run between six and seven feet in height and the new-born, disproportionately tall and spindly, about three feet.

Baby guanacos, chulengos, are born in November and early December, at least in central Patagonia, where this is late spring. They can run almost immediately, although for a day or two their knees knock together and they tire easily. Within a week they are able to keep up with their elders except on a very long hard run. We are told that twins never occur, and this may be true, but twice we saw a single female followed by two chulengos. One may have been an adopted orphan. Once we came across a chulengo of about two months, a time when they are normally still with their mothers, far from any other guanacos. It looked very thin and the sight of our car sent it into a blind panic.

Unique as it is in comparison with all other beasts, the guanaco structure seems to leave little room for great individuality within the species—but probably they think the same of us. Even the difference between males and females is practically impossible to make out in the living animals, except that the males average a little larger and behave differently. There have, however, been famous individual guanacos that were widely known among the settlers. There was one in particular with a gleaming white spot on its forehead. It was an old male that lived near the Cerro Cuadrado, south of the lake basin. The fame of this animal spread so far that a zoo offered five hundred pesos for its capture alive, but it had seen too many hunters to be an easy prey and at last reports it was still free on its native heath. This old Silver-Face was always seen near Cerro Cuadrado, and the natives say that guanacos never go more than a few miles from their birthplaces, soon developing regular haunts. How odd this is in an animal capable of such easy and rapid locomotion, and if their ancestors were equally sedentary, how long it must have taken them to make the long trek down from the plains of North America to Patagonia!

From such known individuals, the span of life has been fairly well estimated, and it seems to be about thirty years, a venerable age for a wild animal. Justino said that he knew one in the Sierra San Bernardo that had been adult when settlers came in twenty-five years ago and that is still alive now. Others reported also that they knew of certain guanacos nearly or quite thirty years old. Among the most widely renowned individuals have been one or two that were albinos, snow-white, but I never heard of a melano, black or unusually dark.

Unlovely and miscellaneous as is his exterior, it is less erratic than a guanaco's mental processes. If actions are a fair guide, his mind is often vacant, sometimes hysterical, and always stupid. It would be charitable to suppose that some of the guanacos that I have seen were insane, and that all suffered from a sort of animal arrested development of the intellect and the emotions. Their psychology, if such apparently vague and disordered thoughts can be dignified by such a term, seems primarily to involve a lifelong conflict between curiosity and timidity.

Among the first guanacos that I saw, and the one that I came to know best of all, was a solitary animal that used to watch me at work on the barranca south of Colhué-Huapí. Safely separated from me by an impassable gorge, he would appear every afternoon and curse at me for intruding in his realm, yammering by the hour. Imagine a tin horse that has been left out in the rain until thoroughly rusty, and then imagine that the tin horse has colic and is trying to whinny, and you will have a faint conception of a guanaco's yammer. The only other sound I heard them utter was an expression of fright and sounded like the first notes of a coloratura donkey, also rusty. Yammering expresses all at once "I see you," "What the hell are you doing on my property?" "Look out, boys, here's something queer!" and "Oh, dear, oh, dear, oh, dear, what *shall* I do about this?" A full translation would also have to include some obscenity, for there is something distinctly indecent about the noise as it issues from the beast's protrusile and derisive lips.

Any unusual thing, such as a man, is sure to attract the attention of every guanaco in the vicinity, and a chorus of yammers results. A guanaco is an artist who simply must express himself. Not for him to take in a situation and then silently slip away. He must get close enough for a good look, and then must express his disgust with the whole arrangement in a full and decisive way. Here is the conflict in guanaco psychology again: when he is afraid of anything he proceeds to call attention to himself. Often we should have been quite unaware that there were any guanacos around if they had not insisted upon telling us so. When we heard a yammer, we could be sure that the beast was in plain sight, perhaps half a mile away, but surely somewhere where he had a good, clear view of us.

This is also demonstrated by the guanaco strategy of retreat. They never hide. When startled, their primary idea is always to keep their presumed enemy in sight. They make for the nearest good lookout point, and if they decide on flight, they run along ridges and in exposed places as much as possible. Perhaps this is not quite so dumb as most of their doings, since it might well be an ideal system against their primeval enemy, the puma, but it is worse than useless against their present archenemy, man. Still, they have only known man for a few generations and their learning ability is practically zero. What was good enough for their ancestors in the Pliocene Epoch is good enough for them.

I have elsewhere described the peculiar gyrations of a flock of guanacos which Justino said, but which I do not believe, were a sort of jubilee at the birth of a chulengo. I suspect that one of them started it because he could not think of anything else to do, and the others followed him because they did not know any better. Certainly they go on climbing sprees. Cliffs seem to have an irresistible fascination for them and they spend hours running up and down barrancas. There is no food for them there, or any other apparent legitimate business, and they seem to climb for fun and because they do not have anything better to do. All the cliffs are covered with their trails, often the most practicable routes for human climbers to follow, and they go almost anywhere that a man can and many places that he should not. Often we have cursed at them as we toiled slowly up an almost vertical trail behind a guanaco who went bounding lightly along, having the time of his life.

[If I let that and some of the other paragraphs here go as they stand, some of my colleagues will accuse me ponderously of teleological interpretations (a dreadful accusation in scientific circles) if not of downright nature-faking. I hasten to add that I am describing exactly what guanacos do and that I do not pretend really to know what they think (if they do think) but merely set down how it looks to a human observer. If the guanacos do not like it, they are free reciprocally to interpret my actions in their own way—as indeed they do, summing it up quite trenchantly in their yammerings.]

They live in a dry country, but they require water. If possible, they drink every day, and they will travel long distances to reach a lake or spring. The guanacos of the dry Valle Hermoso have worn a deep road down to Lake Colhué-Huapí, and they make this arduous journey daily. Elsewhere they will always be found near a spring or waterhole at dawn, scattering to graze later in the day. Even the presence of man within a few yards will not deter them when they are thirsty, as we saw at Cañadón Vaca where they used to come right into camp for their morning drink. On such occasions they came in bands of five to fifteen animals, as a rule, and we usually saw them in such bands even on their grazing grounds during the summer and fall. In the spring most of those we saw were solitary or in twos or threes, but I do not know whether this is really seasonal or whether it is due to our then being in poorer guanaco country or to chance. The bands of moderate size almost always had a sort of guide and sentinel who went ahead to spy out the land and who lagged behind and brought up the rear in retreat. This animal appeared always to be a male, and I believe that the others were his harem, with perhaps some immature males. At all times there were some solitary individuals scattered about, and at least during the later months these seemed to be males—probably disappointed in love. Earlier observers report bands of hundreds or even thousands, but we rarely saw over twenty-five at one time, and the largest band we encountered numbered only forty-five or fifty.

One of the most curious and inexplicable of all the habits of these curious and inexplicable beasts is their way of depositing their dung in certain fixed places. The dung, which resembles that of sheep but is larger, is always deposited on an old pile, and some of these piles still in use must be many years old. How they get started and whether the piles are individual, communal, or used by all comers we were unable to learn.

As a result of another curious habit, hunters new to the country usually think that they hit every guanaco that they shoot at. When startled, as by a shot, and starting to run rapidly, these Patagonian camels plunge, dip their necks, and appear to stumble violently before they hit their stride. After a few preliminary dips, they hold their necks erect, or nearly so, while running. On the flat pampas we several times had opportunities to time guanacos by chasing them with the car until they clearly could not go any faster. The top speed observed was about sixty kilometers per hour, about thirty-five miles, and this only when the beast was making a supreme effort. This is not a remarkable speed for a large plains animal whose sole defense is running. A moderately active dog can catch a guanaco in a race of about a kilometer, and a good horse can pass a guanaco easily in a short run on the flat. In the badlands or in other broken country, however, the guanaco comes into his own and can escape almost any pursuer, for he hardly slackens his pace however rough the going. Ostriches, the other runners of the pampas, are even slower, not exceeding forty kilometers (about twenty-five miles) per hour.

When settlers first came into this country, their sheep provided luxurious fare for the pumas, then very numerous. The big cats gleefully deserted their natural diet of guanaco meat and dined to repletion on the slower and, if possible, still more stupid domestic animals. The result was that the guanacos began to multiply enormously, and as the settlers waged war against the pumas they removed the only real check on this increase. Within a few years of settlement there were probably many more guanacos in Patagonia than there had been in aboriginal days. Then the settlers tried to restore the disturbed balance of nature by hunting the guanacos. Greatly to the detriment of both pumas and guanacos, the balance has now about been restored, on the basis of a much smaller animal population, a sort of revalorization. This, however, has not come about so much through the hunting of adult guanacos as through the once thriving trade in chulengo hides.

Adult guanacos are hard to catch and expensive to shoot. The meat is not good, and the hides are valueless except for the neck which finds a limited use in making plaited leather ropes. Chulengo hides, on the contrary, are highly prized and fetch from one to five pesos apiece. The chulengos are very easy to run down on horseback and catch with boleadoras. The whole outlay for the hunter was his time, not counted as worth anything, and as a result of this invigorating sport he could gain what was to him a small fortune every year. The inevitable result was that hunting became so intensive that in most regions hardly a single chulengo would live to be a week old. In the last five years in central Patagonia it would be a reasonable estimate that each year not one chulengo in ten has escaped the whirling bolas of the chulengueadores.

The species naturally diminished in numbers, but to a casual or a disingenuous observer (and the hunters were always the latter and usually also the former) they seemed to be in no danger of extinction. The real result was not immediately visible. The herds seemed to be holding up in numbers fairly well, but this was because of the longevity of guanacos. Actually, almost no young recruits were growing up, and the greater part of the population consisted of old individuals, fifteen or twenty or more years of age, that were born before the intensive hunting began. As these reached their term of life, they would die off in great numbers and within a year or two, when it was too late to do anything about it, the species would be near the vanishing point. It seemed to me that this crucial period had almost been reached when we were there, and to the great credit of the Argentine officials they saw the danger. While we were still in the field a three-year truce was declared, during which the killing of guanacos (including the young, of course) is prohibited. With such continued watchfulness, there is no doubt that the species can be kept indefinitely.

This is a joyful prospect, for the extinction of Patagonia's guanacos would be a calamity. Properly husbanded, they are an excellent commercial resource, and Patagonia has too few such resources to throw any away, but the real reason for my own rejoicing is frankly sentimental. Patagonia is the guanaco's country. He is better adapted to it than we are and he belongs there. The country and its most typical animal belong together. A Patagonia without its guanacos would be dismal indeed, unthinkable. Ungainly, ugly, and irrational as he is, the guanaco lends to the region much of its life and interest. Long may he survive, sentinel of his native pampas!

He and the ostrich. Unlike as they are, these two seem to go together. Both are ridiculous, unbeautiful, and stupid; both are at home in Patagonia; but there the resemblance ends. Yet to think of one is to think of the other, and any mental image of the high pampas will have both in it.

This ostrich is not an ostrich at all. An ostrich is a large, flightless bird with plumes and with two toes on each foot, and it lives in Africa. These are smaller (but still fairly large), flightless birds without plumes and with three toes on each foot and they live in South America. In books, most books, they are called rheas, but I did not meet anyone in the Argentine who had ever heard of a rhea. The two, "rhea" and true ostrich, differ in many more abstruse ways than in size, plumage, and toes, and some authorities maintain they are not really related at all. Others say that they are; I do not know. In any event it is agreed that they are quite different birds. The rhea, however, is very ostrich-like both in general appearance and in habits, and it is natural that the Europeans should have used the name "ostrich" for the South American bird. That usage is so universal that I will not use the book name, and, except in this paragraph, when I write "ostrich" I mean my Patagonian friends. In Spanish they are "avestruz," pronounced to rhyme with moose, ah-vess-truce', and the natives sometimes also use the Indian name, choique.

There are two species, differing only in minor details. *Rhea darwini*, Darwin's rhea, avestruz patagónico, Patagonian ostrich, or choique lives in the Patagonian plains and in the high Andes of the Argentine. *Rhea americana*, American rhea, northern ostrich, or ñandú (a Guaraní Indian name) lives in the northern pampas. We saw the northern species in the Province of Buenos Aires, but our constant companion in the field was the southern form which was first recognized as distinct by Darwin and the scientific name of which was given in honor of that great naturalist.

The ostrich's head is small, his eyes large, and his expression unvarying and foolish. He looks like a happy but unintentionally malicious half-wit. His ears are quite visible as holes in the sides of his head, surrounded by small feather tufts which do not protrude. The neck is long and gawky and the body much the shape of a battered and outsized football. The wings are large but flabby and are hardly visible in repose, folded back over the body like an overcoat. There is no tail to speak of. The long, coarse feathers of the top and sides of the body and of the upper legs are dirty nondescript grey or greyish brown in color while the finer and shorter feathers of the under parts are white or yellowish. This belly color is generally lighter in the females and in the young, so that hunters follow the rule that the whiter these feathers the better the meat.

The feet are very curious, with a large middle toe and a slightly smaller toe symmetrically placed on each side. They look as if they should not belong to a bird but to some extinct reptile. The track is almost exactly like the trails left in ancient sandstone by the bipedal dinosaurs. When we first traveled into the Lost World of Patagonia, it seemed peculiarly appropriate to find its surface crisscrossed by fresh tracks apparently of dinosaurs.

Although, as I have said, he cannot exceed twenty-five miles an hour, running is the ostrich's long suit. The gait is ludicrous, the feet picked up jerkily and planted heavily in front as if the bird were trying to stamp on something at each step, and as he runs his flabby wings are extended and flap in the breeze. They may be of some use for banking on the turns, but the musculature is so weak that they cannot help very much. There is a ridiculous anomaly in the leg action that baffled us until we suddenly realized that what seems to correspond to the human knee bends in the opposite direction. The bipedal gait seems vaguely human, but it is that of a man whose legs have been put on backwards, an effect due to the fact that the actual knee is high up in the feathered part of the leg and not obvious, while what seems to be the knee is really a special joint between the ankle bones.

As with the guanaco, running is the sole defense of the ostrich, but his strategy is quite different. A guanaco's whole idea is to keep the enemy in sight, while an ostrich is unhappy until he can no longer see his pursuer and then becomes quite contented. A startled guanaco runs for a high place and re-treats along a ridge; a frightened ostrich gets down into nearest hollow and retreats along valleys. Even against his natural enemies, this ostrich system would seem to be almost the worst possible, but it must work for the species survives. Although a pursuer may be close behind, an ostrich will stop running as soon as he has crossed a ridge or gotten down into a little hollow. This peculiar psychology may be the origin of the legend, certainly untrue of the Patagonian birds, at least, that ostriches hide by burying their heads in the sand. Their actual method is not much more sensible.

True to the reputation of the whole tribe, ostriches will eat almost everything, but they are primarily vegetarians. Berries and seeds are most eagerly sought, but during the long periods when these are not abundant, blossoms, leaves, coarse grass, and even twigs and thorns will serve. The young, and possibly the older birds too, although I did not happen to observe it, are inordinately fond of flies and are adept at catching them on the wing.

Like almost all wild creatures, ostriches are always dis-eased. In every one of the many we opened, the intestines were choked with worms, and external parasites are equally abundant. Since this condition is the rule, perhaps it should not be considered as a sort of disease but as a normal ostrich condition, and the animals may be thought of as small migratory worlds, densely populated. Actual skin diseases seem rare in them, unlike the guanacos which often have a very virulent form of scab, very nasty in appearance, from which they eventually go blind—another reason why the sheepgrowers hate them, since the scab can be communicated to the sheep.

Aside from their normal internal and external guests, ostriches have many enemies, especially when young. The hatching season means a continuous field day for skunks, foxes, wildcats, and eagles. The adults are hunted by man, sometimes for food but more often for the feathers. Justino tells me that when he was young (he is now twenty) many of the ostriches used to migrate some miles annually, spending the summers in the sierras, and the winters in the basins and valleys. In spring and fall they could be seen by hundreds on the way, and all the settlers used to gather and slaughter them. Then a few years ago the migration suddenly ceased and now they stay in small groups and in limited areas. I have seen about twenty in a group, but there seemed to be no particular organization and most of them simply go their own way regardless of the others.

Ostrich nests are shallow depressions scratched in the soil. The eggs are about five inches long and are green when first laid, before the sun bleaches them to a dirty cream color. Thirty eggs is an average clutch, but I counted forty-eight in one nest, and was told that as many as seventy-five had been reported. In central Patagonia late October and November is the height of the laying season, and development must take about a month for by the end of November the young were very numerous.

I deeply deprecate ostrich domestic economy, and sincerely hope that humans will not emulate it. The cock ostrich is the housewife, and the hens are flighty creatures who think of nothing but their own pleasure. Each male has a harem of several hens, and he prepares the nest in which they all lay. Occasionally necessity overtakes them elsewhere and it is not

uncommon to come across lone eggs, guachos—a word meaning orphaned, isolated, or alone and not to be confused with
"gauchos" as was done by a lady novelist describing Buenos
Aires. "It is fascinating," she wrote, "to stroll along Florida
and see the picturesque guachos in their native costumes."
—And even a gaucho in costume on Calle Florida would be
a much rarer sight than an Indian in war paint on Broadway.

The hens complete their only racial function, that of laying, in about two weeks. I was unable to check, but Justino
and others say that each hen lays every two or three days.
Then the females forget the affair and wander off together
to the club, or wherever hen ostriches like to go to carouse,
leaving all the work to the downtrodden male. He it is, who
sits on the clutch, covering the outer circle of eggs with his
meager wings, and snatching a little food as best he can
during the heat of the day. He is sensitive and temperamental.
Sometimes if the eggs are disturbed while he is away on a
foraging expedition he will be so upset that he scatters and
breaks the eggs when he returns. There are even some signs
of feeble revolt against his hard lot, as we found some clutches
apparently in perfect order but abandoned. On the whole,
though, he does his duty, hatches the family, and cares for it
until it can fend for itself.

In the hatching season, we used to see the cock ostriches,
each surrounded by a milling mob of little ones. The old
man usually appears quite distracted with his swarming and
active family of twenty or thirty, fluttering and fussing
around them like a domestic hen, behavior very silly and
undignified in a bird almost as big as a man, and male at
that! And what could be more ridiculous than a lost baby
bird as big as a full-grown chicken, crying for its father? It's
a wise bird that knows its own mother. An ostrich Al Jolson
would sing Pappy songs.

Newly hatched ostriches are called charitas, a word that
seems to be purely Patagonian. By the time they are two or
three months old, they have lost their down and grown feathers and are then charas. In a year they are nearly full-grown

and are avestruces. The one-year-old hens usually lay a few eggs, and these may be small or abnormal; the two-year-olds are fully adult. I do not know what the normal span of life is.

We caught several charitas at different times and tried to raise them as camp pets, but our luck was bad, as it was with all our pets. The only one that did not soon escape died an early death of kerosene, as sadly recorded on a previous page from my journal.

While this pet, named simply Charita, lived, it was a constant source of amusement. It was ostrich-like in head and feet, save that the feet were several sizes too large for it, but in between it was like nothing much. Its body had the shape and size of the egg it came from (where the neck and legs fitted in is baffling), covered with very soft down, the back dark brown, striped with white, like the pelt of a skunk.

Charita became quite tame and happy during the short time that we had him. He was the size of a chicken, but this did not deter him from trying constantly to crawl into our pockets. Usually he got just his head and neck in, leaving his scrawny legs waving outside. He also used to crawl into bed with Coley —sleeping with an ostrich was an unanticipated Patagonian experience. Charita loved warmth and showed no discretion in his pursuit of it. He would crawl under the stove between scalding-hot saucepans, and only our constant vigilance kept him from being burned to death. Aside from keeping as warm as possible, his main ambition was to learn to stand on one leg while he scratched his head with the other. He would brace himself well, feet far apart, then lift one leg to scratch and promptly fall down plop! on his beak. Then he would get up, balance himself again with an air that seemed to say "I'm sure I've got the hang of it now!"—and promptly fall on his face again.

Charita's call was a whistle, sliding down the scale and ending with a low, pathetic tremolo. We learned to imitate this, and carried on long conversations with him. He would come running when we called, and when he got lost in the brush around the camp he would whistle and find his way back when we replied. There is, however, a variant of the call which seems to end on two notes whistled at once and which we could not achieve.

Charita had endless and very aimless curiosity. He was into everything in the camp, and went roaming around watching and tasting. We gave him wholesome food, bread, meat, and his choice of the surrounding vegetation, but he tried to eat everything he came across. Adults raised as pets are very troublesome in this way, living up to the ostrich reputation by eating gold watches and any other such tidbits left unprotected. It was this lack of gastronomic discretion that brought Charita's life to an end.

Our luck with pichis was not much better, but in the beginning the trouble was the exact opposite of that which robbed us of our ostrich. The armadillos would not eat anything. We offered them meat, gravy, bread, milk, and everything else that the camp afforded, and still they sulked and starved. Of the first half-dozen or so that we tried to tame, some escaped, others would not eat and so were themselves eaten before they got too thin, and others passed the point of being edible and were released. Finally we got one that ate voraciously and that lived with us for a long time, only to die a tragic death when we were on our way back north.

The way in which this pichi was persuaded to eat was not pretty, but was effective. We had tempted her appetite with everything we could think of, and still she refused and was pining away like the others. There was some ostrich meat that we had kept too long and that had finally been thrown away some distance from camp. It had lain in the sun for about two weeks, and it was exceedingly foul, but in desperation we cut off a chunk and offered it to Florrie. To her it was nectar and ambrosia. Her apathy disappeared as if by magic, and she tore into that putrid flesh eagerly and soon made up for her long starvation. Thereafter we never had any trouble getting her to eat, and by slow degrees we weaned her away from her perverted taste until finally she would eat almost anything, even fresh meat.

All our camp armadillos were named Florrie, by a pun too complicated to be funny. Pichi is pronounced Peachy. "Flor de Durazno" means peach blossom, and is also the title of an Argentine novel which we had read. We called the first pichi Flor de Durazno, and soon shortened this to the less cumbersome Florrie. *The* Florrie was the one that lived with us for so long, and the others were mere transients who borrowed the name for a while. "Pichi" is an Araucanian Indian word meaning "little" and it is the common colloquial name for one of the species of armadillo that lives in Patagonia, known to science as *Zaëdius pichyi*, because it happens to be smaller than the more northern peludo, or hairy armadillo, which it otherwise closely resembles.

Adult pichis reach about eighteen inches in length, of which a fifth is head, a half body, and the rest tail; the neck is so short and so hidden beneath the shield that it does not count. The head is conical, with a small, round, very mobile, rather pig-like nose and a long mouth, extending back beneath the eye, with very flexible lips but no true cheeks. The ears are small and hardly break the outline of the head, unlike another northern species that is called muleta, little mule, because of its big ears. On top of the head is a triangular armor plate made up of small scutes. The top and sides of the body are also covered by the armor for which armadillos are famous. This body armor comprises a slightly flexible but coherent shield over the shoulders, another more than twice as large over the hips, and a series of seven movable rings between the two shields, the rings overlapping like shingles. The tail is completely enclosed in from six to nine rings of armor.

The under side of the animal is bare except for sparse, coarse, long hairs. Even the armored part is somewhat hairy, for bristles grow from the hind ends of the separate scutes, dark brown with white tips. The scales themselves are a dirty brown, tending to a lighter cream color along the edges of the scales and near the margins of the armor. This armor consists of many little flat bones embedded in the skin and of still more numerous horny scales overlying them. It is very tough but is not really hard, even the fixed shields being flexible when pressed.

Pichis are very bowlegged, their legs short but strong, their feet stubby and provided with large, sharp claws which are longer on the forefeet. With these they can dig with amazing rapidity. A man can run a pichi down with no real effort, but if the pichi succeeds in reaching soft ground, preferably beneath a thorn bush, he is underground in a moment and practically impossible to get out again. If overtaken on ground too hard for rapid digging, the brute takes firm grip with his claws, pulls the edge of his shell down all around so that it is very difficult to get a purchase on it, and holds on.

This was a great disappointment to us. We had learned, as have most people, that armadillos defend themselves by curling up into a ball. In spite of his numerous rings, a pichi cannot come as close to curling into a ball as I can, which is not very close. He can only hunch himself up a little bit, and this he does to keep warm, or when resting, and not as a means of defense. Of the many kinds of armadillos there is just one, not a native of Patagonia, that can curl into a real ball, but popular imagination has extended this ability to all his brethren.

When severely frightened, pichis express their emotions by violent defecation, and this makes a newly caught animal a very unpleasant companion. Their armor is insensitive, but the hairs protruding from it are very delicate sense organs. If a sleeping or quiescent pichi has one of these hairs barely touched it will straighten out with a jerk, emitting an explosive wheeze, a very disconcerting maneuver. Otherwise it makes no serious attempt to defend itself. There are no teeth in the front of the feeble jaws, and those in back are mere blunt stubs.

Florrie became quite tame and soon learned to tolerate our attentions without any reaction except for a little squirming. Her tameness, however, was not based on affection or even on the realization that we were an easy source of food. She came to accept us as a necessary evil, and as nearly as we would let her she would ignore us and go about her affairs. These affairs consisted largely of trotting about from bush to bush, pushing her nose into the earth, and sniffing violently. Life to Florrie was a succession of smells. She seemed to pay practically no attention to anything she saw or heard but lived entirely by and for her nose.

Florrie developed a passion for condensed milk. She would lunge into a saucer of it, getting both front feet into it and often spilling it in her piggish haste. She lapped it up with her long, prehensile tongue, but not content to do this neatly and efficiently, she would plunge her whole nose in eagerly, with a mighty blowing of bubbles, and then choke and retire wheezing. Much as she loved it, she was so stupid and vague that she would completely forget where it was if she got a foot or two away from it, and then would wander around aimlessly for a while. If she happened to come across the saucer again in the course of her rambles, she would once more set to, as if she had never been away, and once more there would be bubbles and wheezings.

Once she wandered away from camp and was not seen for several days. We mourned her for lost, but one morning there she was again. I am sure that no affection nor even any glimmer of memory brought her back. She was too vague by far for that. She had simply been wandering about the vicinity and happened to wander in on us again. For sheer vacuity it would be hard to match an armadillo. Even a sheep shows more sense, and the clownish guanacos are geniuses in comparison.

This makes it rather hard to feel any real affection for the beasts and there is no more inducement to pet them than if they were turtles. Yet we did become attached to Florrie for all her dumbness and self-absorption, and we sincerely mourned her passing. Poor Florrie! On a rough road one day down in Santa Cruz a heavy wagon-jack was jolted onto the box in which she was riding, broke it, and crushed her. Her skin was not quite thick enough.

We had several other pichis during the latter part of our

stay in Patagonia, four at one time including Florrie, and we made a great effort to bring one or more back with us, but one by one they died, or escaped, or were stolen, and in the end we returned pichiless. They were our most constant camp companions and amusement, and we were seldom without one or several as long as we were in camp.

We were not at all successful with other pets. Guanacos are quite easily tamed, but the problem of transport when we moved camp and later when we were constantly traveling was too great for us to attempt it. They are not very pleasant companions, anyway, for when domesticated they lose all their timidity and retain all their less fortunate characteristics, including that, apparently common to all sorts of camels, of spitting expertly in the faces of human foes and friends alike without warning. The baby owl that I have already mentioned never abated his virulent hatred for us, and one day he struggled so hard in one of his transports of rage that he broke the leg by which he was tied, and so had to be killed.

We never tried taming martinetas, finding them too good to eat and too hard to catch alive. These are partridge-like birds, speckled brown, black, and white, nearly as large as domestic fowls, actually a sort of tinamou, although "tinamou" is a book name like "rhea," that is not known to those best acquainted with the birds. The distinguishing features of the martineta, as compared with other tinamous, are a feather tuft or crest on the head and its relatively large size.

This martineta is a curious bird that can fly but does not like to. It normally stays on the ground all the time, and runs when startled. Only in extremity will it fly, and then low, heavily, and only for about a hundred yards. Once it has flown, it usually cannot be persuaded to do so again but will "throw itself," as they say in Spanish, that is, squat down and remain absolutely motionless. Then its protective coloration comes into play and it is extremely difficult to see. More than once I have almost stepped on one before I saw it. The local inhabitants take advantage of these habits in hunting the martinetas—another example of a defense excellent against animal enemies but peculiarly disadvantageous against man. Once the birds have flown, the hunter will note where they light, circle around the spot two or three times until the whole covey has thrown itself, and then seek them out and knock them on the head. This reluctance to fly is the more strange because the birds have large wings and the breast muscles are powerful.

The most peculiar thing about the martineta, however, seems to me to be its resemblance to the ostriches. In size and build it is very different, yet the points of similarity are numerous and important. The feet, for instance, are exactly like ostrich feet (the three-toed, rhea kind, of course) in miniature and the tracks are like those of charitas although still smaller. The preference for running is also a point of similarity, and seems graphically to illustrate the first stage in the development of a flightless bird. Only a little more emphasis on this habit, and the martinetas would cease to fly altogether. Then their wings would quickly degenerate and the mass of breast muscles disappear, and within a few generations the possibility of ever flying again would be lost. And once they were flightless, there would almost surely be a steady increase in size.

I predict that the martinetas of a few hundred thousand years from now, or a million or two, if they survive that long, will be ostrich-like in every way. Or, put in another way, it seems highly probable that the ancestors of the Patagonian ostrich in the remote past were very martineta-like. Quite possibly the ostrich and the martineta are actually relatives, the ostrich merely being a little more advanced in its evolution.

Just why the martineta is becoming flightless and the ostrich has become so is mysterious, but encourages speculation. Both seek their food on the ground and perforce nest there in the total absence of trees in their homeland. Both seem to have found in running a defense which is adequate from the point of view of race-survival, however fatal it may be to many individuals. Although there are such obvious objections that it cannot be advanced as a real theory, I suggest that the wind may have something to do with it. It is very difficult to fly in the wind. Eagles take advantage of it and soar, but a non-soaring bird which eats and lives on the ground might well find that flight was positively disadvantageous here. Taking off and landing in the wind is very difficult for birds of any size and once in the air it is often impossible to fly where they want to go. This was strikingly seen on the numerous occasions when we saw wild geese desperately trying to fly westward and being carried back to the east despite their best efforts, while on the ground the ostriches were able to run westward with great ease.

The resemblance of the martinetas to the ostrich even extends to their family life, so I was assured by all the local inhabitants, although unable to establish it on my own authority. They say that the martinetas are polygamous and that the male covers the eggs and raises the young.

The nests, placed on the ground under thorn bushes, are made of twigs and other small vegetation and lined with feathers. The eggs, nine or ten in the cases where we could count them, are slightly smaller than hen's eggs and not quite so elongate. The surface has a high polish, like finely glazed porcelain, and the color is brilliant green, verging toward olive. The young can fly while still very small, and are somewhat quicker to do so than are their elders. Several times I came across families and the old man threw himself at once while the babies positively exploded into flight in all directions.

Martinetas prefer brush-covered country and are most active at night. They are often encountered at dawn or at sunset, but during the bright day are usually hidden under bushes. Sleeping out in martineta country, one can hear them whistling on all sides and all night long.

Their cousin the perdiz colorada lives in the same region, but the two species are seldom seen at the same time. The perdiz colorada prefers more open country, of the type of the Pampa Pelada, with little or no brush, and it is more likely to be active during the day. "Perdiz colorada" means "red partridge," but like "avestruz" this is a misnomer as the bird is quite unrelated to the true partridge and is a tinamou. As the name implies, it is somewhat more russet in color than the martineta; and it is also a little smaller and lacks the crest.

Both martineta and perdiz colorada are very good eating, although rather dry. For a time, where they were particularly plentiful, we had so many that I became quite sick of them and gladly went back to a pure mutton diet, although Coley continued to relish them. Near the towns they are caught in large numbers by setting up low nets and driving the birds into them. In the provinces there is a still smaller tinamou, about quail-size, that is always to be found on the market in Buenos Aires and that is even shipped out in large numbers, frozen, and sold as partridge, to which it is not inferior in flavor. These little tinamous may often be seen alive in the windows of butcher shops in the cities. In Patagonia, however, the tinamou tribe is represented only by the martineta and the perdiz colorada.

Another delicacy, but one less appreciated, is the liebre patagónica, or Patagonian hare. Here again the European settlers have applied a name from the Old Country to a creature of wholly different relationships in the New, and here they have gone even farther astray than usual, for the liebre patagónica is not a hare and except for its large ears does not look like one. It is a rodent of the guinea-pig family, although the external resemblance to a guinea-pig is even less than to a hare. It is large for a rodent, about twice the size of a jack-rabbit, and looks peculiarly like a little donkey except for its more elevated rump. The rump is black, the rest of the body tawny brown and white. The tail is short, but not woolly.

The liebre patagónica has a gait such as I have never seen in another animal. It jumps, all four legs almost at the same time, and very stiff-legged, as if it were bouncing along on springs, or, as Coley remarked, as if it were mounted on pogo sticks. This makes it a very difficult target, and Coley became quite exasperated before he succeeded in bringing one down—and then discovered that his single shot had killed two of them! This was due to another typical habit of theirs, that of living in couples, male and female, which rarely are more than a foot or so from each other. They run together and stop together as if they were Siamese twins. Their homes are large, deep burrows.

Liebres patagónicas are rare, although little hunted by man; the fur is not valued and the meat, although good, is seldom eaten. Their present scarcity followed and is undoubtedly due to the introduction of European hares in Patagonia. These eat the same food as their Patagonian namesakes and are much more prolific, so that they are running the native animals out, which I think is tragic. The European hares are quite worthless and do not belong in Patagonia anyway. They should never have been introduced.

There are two kinds of wildcats, gato montés and gato pajero, both unlike our familiar wildcats, but we did not happen to encounter any. The only other cat is the puma, which belongs to a local species but is almost exactly like the North American cougar or mountain lion. The puma is probably the most adaptable of all animals. With only slight local variations, it ranges from Arctic North America down to Tierra del Fuego, on mountain tops and plains, in dry country and wet, treeless plains and tropical jungles. Then there are two sorts of foxes, or so-called foxes, for they belong to peculiar native genera and have little in common with the true fox beyond belonging to the general dog family. Skunks are numerous and resemble our striped skunk both in odor and in appearance except that the fur is not really black but a still more attractive rich brown. We had an all too good example of the pervasiveness and persistence of their perfume. While at Cañadón Vaca, Coley shot one as a specimen for the Museum and, not wishing to skin it on the spot, chucked the body into the back of the truck where it was for about half an hour, until we got to camp. Later we washed the truck body with water, rubbed dirt where the carcass had lain, and finally in desperation even sprinkled gasoline and burned it. About a year later when we sold the truck in New York, it still smelled distinctly of skunk.

Like their northern relatives, the Patagonian skunks are nocturnal and are seldom about in the daytime, so that at first we thought that they were rare. When we had been about more at night, we found that they are very abundant. One night walking into camp in the dark Coley saw something moving and picked it up thinking it was a pichi. It was a skunk, but the story has an unexpectedly happy ending for he discovered his error in time.

There are also weasels or ferrets, unusually dark in color, but we only saw one pair and that from a distance, and we did not see any otters although they are said to occur.

These and many others, some of which are mentioned casually in the pages of the narrative, are the beasts and birds of the pampas. Along the shore there is a whole new animal society in which the seals and the penguins are my favorites. We later spent some time traveling along the coast, and there will be occasion to tell more of its life when I come to that part of the story.

Such is Patagonia's family, a rowdy lot on the whole, but fascinating. The feeling of desolation and of lifelessness is only a human impression and mood. Actually, the land is swarming with life, and to its true children Patagonia is not a desolation but a rich motherland.

CHAPTER XI

THE END OF CAMP THREE, AND OF BALIÑA

Camp 3,
Cañadón Vaca
Jan. 10.

YESTERDAY we called on the government geologist, Piátnitzky, who has a camp on the other side of the river, to see
his territory there and to ask him to visit us and show us a
fossil locality he mentioned some time ago. Like other government camps, his is so luxurious that it is hard to believe
that any work is done from it. There are separate sleeping,
working, and storage tents, each with wooden floor, and there
is a separate sheet-iron cook shack. They also have a regular
menagerie: ducks, chickens, turkeys, goats, dogs, cats, and
horses. And they transport food for all these as well as for
the men.

We slept there last night and this morning came to camp
across the new concrete bridge and then along a trail up on
the pampa which we had not traversed before. We went
astray and got down into a cañadón where the trail dwindled
and died. No one had been there for years. There was an old
abandoned house with one staring window. The roof had
fallen in. There was also a long dead horse. Nothing else
but untracked sand dunes. The place was like a nightmare.
We got out of the sand with some difficulty and finally to
Niemenhuis's, where we found that the smallest girl had fallen
off a cliff and broken her arm. And so to camp, where the
cook greeted us like long-lost brothers. He was down to his
last quarter of mutton and had visions of either starving to
death or having to walk a few miles; of these two dreadful
alternatives he would prefer starving.

Nights are still cold, but during the day the midsummer sun is warm. Coming across the pampa (nameless, as far as I know) which lies along the Chico southwest of Cañadón Vaca and the Pampa Pelada, there was a good mirage, which is an indication of warmer days as I have never seen one in really cold weather. Cerro Salpú, rising above the pampa, seemed to be an island reflected in a broad lake. It will be too small to show at all well, but I photographed it to prove the point, still curiously doubted by some, that a mirage is not a mental delusion but an actual reflection in the air which would look the same to anyone, or to a camera. We also picked up a young pichi on the pampa, and will have another try at taming one.

[This was our great success, our Florrie, mentioned in the last chapter.]

Jan. 15.

The even tenor of our days of work at the barranca was interrupted today by the promised visit from Piátnitzky, who is spending the night. He visited our diggings and was much interested, but like others here he was shocked that we do the work ourselves and cannot see why we do not have peons to do it while we direct them. He has not the usual Russian gift of tongues. His Spanish is elaborately correct but deliberate, and he visibly searches for words, more accurate than would be hasty speech, when he does find them, but not colloquial. Now he is studying English and can read quite a lot and painfully articulate a little, but he cannot understand a word, not the commonest word spoken very slowly and distinctly. It is a curious sort of deafness; the sound of a word which he knows and can say himself simply does not make any connection or sense in his mind when someone else says it.

It is perhaps his meticulous but terribly ponderous habit in Spanish, our sole means of communication, that makes him so unsatisfactory as an informant or even as a conversationalist. He is always flirting around the borders of ideas, so that you see vaguely what he may be driving at, but never quite bringing them out into the light, illumining them, and making them indubitable. He is ineffectual, certainly in speech and probably also in nature. This is not to say that I dislike him. On the contrary, he is a man of singular charm, unassuming, cultured, friendly, helpful, and I am delighted to know him and to have him here.

His slow seriousness awakens some imp of perversity. There is a plant here the name of which sounds as if it might be spelled quilimbay, a thorny shrub, a foot or so high, with bright yellow but minuscule flowers, thorns, and small glossy leaves each of which ends in a single spine similar to the multiple spines of holly leaves. These leaves have a pervasive bitterness the like of which I have never encountered. Holding one in the mouth for a minute is a condemnation to being haunted by the flavor for many hours, and no amount of rinsing or of eating other things hastens its departure. I jokingly offered one to Piátnitzky, supposing of course that his years down here had taught him to beware of them, and asked if he were not acquainted with their sterling medicinal qualities. He was not, and he ate the leaf. He is still going around very unhappily, endeavoring to control the puckering of his lips in order not to betray to me his rudeness in not enjoying the treat that I recommended to him.

Again, at the barranca today, Coley called out and asked me whether I had found anything, and I replied that I had found ten dinosaur skeletons (there being no dinosaurs whatever in these beds). Our visitor came trotting up in a state, as nearly beside himself as he ever gets, wishing to see the skeletons. I explained that it was a joke. "But you *said* there were skeletons!" His only conclusion was that North Americans habitually lie to one another, for some reason which he is willing to believe inoffensive but which is entirely beyond his power to grasp.

Jan. 16.

Today we went off with Piátnitzky, crossing the Río Chico at Paso Niemann where there is a wooden bridge with actual, even though self-made, roads leading to it. We looked over the ground in a new large area of badlands, Cañadón Hondo, and then, late afternoon, Piátnitzky went his way back to his camp at Cerro Tortuga and we to ours at Cañadón Vaca.

We stopped at the bridge where there is a boliche run by one Nollmann and his wife. Our fame is spreading. There was a couple there from the other side of Sierra Chaira, which is hinterland to a country that is all hinterland, and they told us with roars of laughter that some stupid neighbor had told them that we had brought our truck all the way from North America and that we were hunting for old bones. Our solemn assurance that this is indeed true nearly shattered their faith in common sense and reality.

We also stopped in on Duvenage for a moment, his puesto being beside the road. Along with a tedious account of his personal health, regarding which we are already only too well informed, he touched lightly on weapons and robbers. Social status here, he says, is in inverse proportion to armament. Gentlemen carry the ubiquitous cuchillo, a sort of butcher knife stuck in the back of the sash, and perhaps a revolver, while if you see a man who also carries a two-edged dagger, a second revolver, and two rifles, you may be sure that he is scum, a bushranger.

He also told of a puestero, a gentleman by this definition, who was troubled with horsethieves. When a horse disappeared, he usually buckled on his revolver and departed, to return a day or a week later with the horse and without the thieves. Once, to vary the monotony, he caught two thieves and instead of shooting them tied them tightly back to back so that neither could walk without carrying the other. In this manner they proceeded about eight leagues, some twenty-five miles, without food or water, before they found someone to turn them loose. They reformed.

These were Chileans. "There ain't a Chileno that ain't a robber," says upright Duvenage, "and I've known thousands of them." Without mentioning the tale of his shotgun wound, we suggested that Timoteo Hernández seemed to be honest though Chilean, but Duvenage says that Justino's father once sold to Duvenage a sheep that turned out to belong to someone else and created some hard feelings.

Jan. 20.

A typical day that may be taken as an emblem and sample of all unrecorded days in the field at work.

Daybreak. The cook's voice outside my tent.

"Doctor! Ha salido el solcito!"—The sun is up!

I, very sleepy.

"Well, let it go back again!"—A daily formula now so familiar that it draws no chuckle but would be missed if omitted.

Dressing is simple: shirt, trousers, alpargatas (canvas slippers with rope soles), and ready. At the cook tent,

"Good morning, cook. Nice day."

"Yes, but there is the damned wind."

"As always, caramba! What's new?" We keep up this pretense of being in constant and close touch with the pulse of the world. Baliña produces a paper dated Bahía Blanca, December 29th. This is something of a feat. How it got here in three weeks is mysterious. He indicates an article which he says is interesting, reopening the never fully solved question of whether or not he can read. He adroitly escapes my cross-examination on the contents of the recommended article. The report is of a conversation between three old dames discussing the shocking clothes of today and predicting retrogression to the fig-leaf of Eden. The author then abandons his spokeswomen and launches into a long, heavily facetious, and highly indelicate discussion of the relative merits of fig and grape leaves as clothing. Meanwhile Coley has appeared and we share the maté.

"Gracias!"

"Buen provecho!"

There is hot water for washing. Baliña grumbled a great deal at first at having to provide this luxury, and still is resigned in a surly way and disapproving. My toilet done, I write in my notebook while Coley takes his turn at the bottom of a kerosene tin which serves as a washbasin, and while breakfast is cooking.

Breakfast: slabs of mutton fried in oil, with a bit of salt pork, sweet chile peppers, onions, and garlic to make it palatable, a few tortas fritas, and coffee. We change to heavier shoes. The cook has put our lunch in the car. We lurch up our private road to the pampa.

The male guanaco with seven wives is off to the left and that with five nearer the road to the right. When we came one had eight and the other four; guanacos seem to be fickle too.

Then we drop down into the cañadón and make the long traverse of its broken land. At Van Wyk's the dog sheared like a lion plays the traditional game of Showing-them-off-the-place, running down the road when he hears us, sitting until we arrive and then darting back and forth in front of the truck, barking furiously, until we are off his property. In the doorway of the hut the bow-legged girl holds the little white dog to keep him from joining in the game. She nods, but does not smile, being modest withal and warned against city slickers, although she would probably be furiously disappointed if she knew how safe she is from any attentions on our part.

The ascent to the Pampa Pelada is a laborious business and the radiator of the car boils. We stop at Duvenage's fence to give it fresh water from our built-in storage tank, then strike off cross country to the top of the barranca. There we stop and load ourselves with knapsacks, tools, flour, canteens of water, extra shellac, and lunch. A pause to look over the cañadón.

The view is stupendous, one of the most magnificent things I have ever seen. In the foreground is the great cañadón, an enormous gash in the earth, hundreds of feet deep and about five miles across. In it is a treeless jungle of badlands, peaks, ridges, and minor valleys carved into the dazzling white ash and the brick-red sandstone. Beyond it rise the pampas, step by step, like broad stairs rising from hell, but not to heaven. Sharp across the cañadón is the jagged mass of Cerro Salpú, and in the far distance the black bulk of Pico Oneto. Scattered along the horizon are numerous lesser peaks, looking exactly like volcanoes (which they are not). Above, in the smooth bowl of the sky, eagles soar and below, in the tangled shapes of earth, lions roam.

Today is especially vivid. Numerous local storms sweep across the landscape, the clouds are bright yellow or deep purple as the sun strikes them or not. The peaks disappear and then reappear as the falling rain passes before them, and occasionally a distant barranca gleams white in the sun and then merges again into the general blue of the horizon. Nearby, the sunlit falling rain is like a trailing veil of luminous gauze. To the east, above the incredibly flat surface of the Pampa Pelada, arises a brilliant and perfect rainbow. The scene is thrillingly, deeply beautiful and its fundamental harshness seems merely the strength and character that differentiates beauty from prettiness. There is nothing pretty in Patagonia, even at its best.

With all of this Coley heartily disagrees. To him, this view is merely a lot at once of what is not worth seeing even in smaller amounts. He likes placidity, trees, water, hedgerows. Probably he is right; I do not scorn those gentler charms, but to me there is something unutterably fascinating in desert landscapes, even when they are wholly bitter, which this is not. We do not argue the point. We are accustomed to the fact that some like water while some like deserts, and that he is one while I am the other. I have learned to respect Patagonia; transiently I seem to hate it, but underneath I still like it. Coley still does not like it and would never learn to accept or, I feel, to appreciate it.

We go sliding down the steep upper barranca, covered with a talus of pebbles from the pampa and worthless for fossil hunting, and at the foot of its two hundred odd feet of descent we come out on a flat bench which we cross. Then for about two hundred feet more we descend diagonally across the badlands proper, along precarious guanaco trails and gingerly down crevices in the vertical beds of compact ash. At their base, these beds of ash rest on a still thicker but fossilless series of grey clays and red sandstones, extending down into the utmost depths of the cañadón. On the lowest ash beds we set down our burdens and start to work.

Duvenage calls this place La Oficina del Diablo, The Devil's Office, an unimaginative name but one that will do as well as another. It is a natural amphitheater cut into the barranca, an amphitheater with a seating capacity, I might guess, of about five hundred thousand people. At present it is occupied by two (granted that flies are not people). In the center is a little conical hill, about a hundred feet high, down the sides of which are spilled streams of brown pebbles. It looks like a Brobdingnagian serving of ice cream, with chocolate syrup. An elevated runway on a hard rock stratum joins it to the main barranca, and along the latter the same stratum forms an aisle, convenient for prospectors. On one side of the amphitheater is a small, deep zanjón, a tangle of thorn bushes through which runs a trickle of water from a spring. Springs are not uncommon on the pampa rim, but only a few have potable water and none produces a flow large enough to trickle away for more than a few hundred yards at most. On the other side there is a high pass over the projecting spur from the cliff, and beyond this is another but less perfect amphitheater, a sort of annex.

Both amphitheaters are extraordinarily rich in fossils, which occur chiefly at three levels, all near the base of the ash beds. Each of us is working on several specimens at once, tending each in rotation and passing on to the next to give time for the shellac or bandages to dry. We have two prize specimens. One, now practically done and ready to haul away, has turned out to be most of the skeleton of *Albertogaudrya*. Ameghino discovered the animal, and gave it its double-barreled name in honor of his distinguished French colleague Gaudry, but he had only a few teeth, and could not envision the real character and appearance of the beast. Now we have nearly complete skulls and jaws, and to crown our joy this skeleton, imperfect, to be sure, but enough to show what the animal really looked like, perhaps even enough to make a restoration and mount the skeleton. It is far the oldest fossil mammal skeleton yet discovered in South America. [*Albertogaudrya* is the largest of the homalodontotheres, summarily described in Chapter IV, from the Casamayor Formation.]

The other prize is even more unusual. For sustained interest and even high excitement, the necessarily slow and partial uncovering of such a specimen is unequaled. First discovery: "Here's a fossil, and it's either a big one or a lot of a small one!" Preliminary examination: it is a string of vertebræ and ribs of some sort, perhaps another *Albertogaudrya* skeleton. As work progresses, it becomes obvious that this is not *Albertogaudrya* but something new and strange. The exposed end of a vertebra finally shows beyond doubt that it is a reptile. The chances favor its being a crocodile, as fragments of these are common in this formation, but further work shows this theory to be mistaken. What can it be—reptile, but not crocodile, with many vertebræ, all with ribs, and yet no sign of legs? It must be a snake! I had not thought of this sooner, because snakes are among the rarest of fossils (I never found one before) and this would have to be the great-granddaddy of all snakes from its relatively huge size. [It was a snake, a new genus and species, and one of the largest snakes, living or fossil, ever recorded, if not the largest.]

At noon, one of the local showers passes over us, and we crawl into a shallow cave and eat while water cascades over the ledge above us. Fortunately it does not rain long, and we can return to our specimens. At six, however, the rain starts again and this time with a determination that shows the hopelessness of trying to accomplish any more today, so we toil back up the cliff, now very slippery in places, and arrive at the car with our feet weighing many extra pounds from the caked mud.

The car slides down the zanjón and swerves perilously past

Van Wyk's. In the bottom of the cañadón we get stuck, but our technique on such occasions is now almost automatic and we are on our way again in only fifteen or twenty minutes. Back at camp the maté goes the rounds, and in the intervals we get washed up and put on some dry clothes. For dinner there is the end of a somewhat dried-out salami, a luscious cauldron of mutton broth, and a meat course of boiled mutton—we cannot entirely break the cook of his partiality for the tasteless puchero, but after all it does produce wonderful soup as a by-product, the ways of cooking mutton are limited, and we cannot expect all to be equally tasty.

After dinner we sit and talk for a while and I write up field notes and journal. Everyone is in good spirits: Baliña because we had read to him from an old newspaper that unusual storms in Europe had killed many people; Coley because he has mastered the art of rolling cigarettes satisfactorily; and I because of the fossil snake. At about ten we all retire to our several tents.

For a time I lie awake, vaguely oppressed by the melancholy thoughts that sometimes take advantage of physical exhaustion after the restraint of action has been relaxed. Home is so far away. A physical weakness that I have not been able to conquer keeps me in constant pain. Life here is not a pleasure in the ordinary sense of the word—but it has its compensations—yes, on the whole I am happy to be here—I do like it in spite of everything—and so drifting off to sleep.

Comodoro Rivadavia
Jan. 25.

I woke before Baliña this morning, an unusual event, and for some minutes lay in bed watching Harry the Hypocrite. Harry is a praying mantis who shares my tent with me, he occupying the ceiling and I the floor. For seemingly interminable periods he will stay perfectly motionless, his long front legs folded up under his chin as if supplicating his Maker to have pity on the sinful world. But if a fly settles down near him, he forgets his prayers and reveals his true nature. The legs unfold and pounce like lightning, landing on his hapless fellow insect with a resounding thump. The fly is consumed with gusto, and then Harry resumes his hypocritical attitude and his deceptive calm.

After preparations and various chores, we left for Comodoro in search of necessary supplies. Here is a condensed version of our log:

Time	Kilometers from Camp	Remarks
11:10	0.0	Leave camp.
11:15–11:35	3.1	Kill and clean ostrich (for presentation to Szlapelisz).
11:53	—	Shoot at guanaco's tail (they do not run from us any more and this is Coley's way of keeping them up to snuff).
12:02–12:05	12.7	Stop at Van Wyk's (he wants us to buy him some chewing tobacco in town).
12:12–12:14	14.1	Stop at gate to water man and car.
12:15–12:34	14.3	Stop at Pampa Pelada to hunt and clean perdiz colorada (Coley making a perfect score of four shots, four birds).
12:47–12:48	22.7	Duvenage's.
1:45–2:54	50.2	Lunch at Nollmann's boliche.
4:16	84.5	Join main Comodoro-Trelew road.
4:22–4:29	87.9	Stop to fix chain (road very bad from recent heavy rain).
4:37	91.8	Start down Cañadón Ferrais (descent from Pampa Castillo to the sea).
5:33	128.3	Arrive at Hotel Colón, Comodoro.

And so to bathe and change and then to dine with friends at Manantial Rosales.

Camp 3
Jan. 29.

We were fiercely determined to get back to camp two or three days ago, but did not make it until late last night. The delay was due to some trouble about a shotgun, some more trouble about a borrowed car, time spent getting ourselves in jail and still more getting ourselves out again, and so on. When we did succeed in tearing ourselves away from the guardians of the law, Tobin came with us.

Tobin is an American and he sells oil well supplies, a business practically cornered by our fellow-countrymen. His office is in Buenos Aires, but he had to come down to Comodoro where the actual wells are to clinch things. (Incidentally, he is not the effulgent liar who came down with me on the *Frers*. Tobin is not a liar, or, at least, I will not admit that he is any better than the rest of us at lying. The effulgent one, I have since heard, was found to have put the support of two mistresses on his expense account, along with the cost of some very snappy jewelry for these and other ladies in Buenos Aires and elsewhere, and this strained his relations with his company. Finally he had to ship one of his young ladies and her mother back to the Old Country, for various and not wholly creditable reasons, and he made the mistake of also putting this in as an expense in selling derricks. This was just a little too much, and he has severed his connection with the firm and disappeared from his accustomed haunts.)

In Comodoro, Tobin met El Rey, Scottie, Whiskey-proof, and the rest. He found Comodoro both quaint and desolate, as it certainly is, and its inhabitants seemed to him fascinating characters, as indeed they are. He imagined the Patagonian hinterland as being much like his familiar old territory of Oklahoma, which it is not in the least, and we offered to take him out to camp for a while so that he could see for himself.

He is far from being a fool, and he has knocked around the world a great deal, but he had never been in anything like a Patagonian camp before, and he departed hurriedly today (in Scottie's car), chastened and aghast. To us it was a curious glimpse of revised values. From custom Patagonia has become to me quite a livable place and our camp, if not luxurious, still a home from home and a very satisfactory place to be. Seen through Tobin's unaccustomed eyes, the bleakness of the pampas suddenly reappeared with full force and the camp seemed outlandish and horridly uncomfortable.

We reached camp after midnight, and this was an inauspicious introduction for anyone who had never been there before. It is a weird sensation to drive along unbelievably bad roads, or rather tracks, circled by darkness, finally to lurch down a steep slope that seems to end in the nethermost pit, and to come upon our little canvas town. We were unspeakably tired and so cold we could hardly move. Baliña appeared, his face in the light of a candle looking more than ever like that of a born cutthroat.

Seeing that Tobin was almost knocked flat by the strangeness and the wildness of his new experience, we could not help laying it on a little for him. He slept in Coley's tent and noticed a revolver hanging on C.'s cot. "Oh, yes," said Coley, "I always keep it handy. Things are always coming in during the night. If it isn't rats it's something worse, so I just shoot at anything I hear moving!" Thus our poor guest was robbed of the little sleep that he might otherwise have had and that he needed so badly. Then in the morning we deprived him of his appetite for breakfast by giving him ostrich meat and by describing quite correctly but in cruel detail the proper procedure in eating sheep's eyes—gouging them out of the socket with a small spoon.

So he left, very soon and unhappily. I should feel conscience-stricken (for I like him), except that his discomfort will soon wear off and he will have a weird, savage, and unique experience to embroider on and to bore people with for the rest of his life.

Feb. 4.

After several peaceful days of the usual round of work at the barranca, we suddenly found ourselves this morning faced by an acute and very dangerous crisis.

Baliña called me early, purposely neglecting to call Coley. After I had washed and had started to take maté, he looked up sullenly and said with amazing calm,

"Señor Doctor, I have decided to kill Señor Williams."

I thought it was a joke, laughed, and handed back the gourd for another filling.

"No, doctor. I cannot stand it any longer. I am going to kill him today. I did not want you to be disturbed when I kill him, so I thought that I would tell you about it this morning, and then I will go over and cut his throat in his tent."

It was no joke. He was deadly serious. Yet he was so calm about it, that I decided to talk to him instead of making immediate plans to overpower him.

"But, Baliña, that is a serious thing and you have no good reason."

Then the floodgates were opened, and there poured forth a torrent of silly little grudges, fancied slights, insignificant incidents over which he had silently brooded until in his disordered mind they had become enormous wrongs which could only be wiped out with blood. The most recurrent theme, and perhaps the one thing that had started the whole fixation, was Coley's dislike of garlic. Baliña is a little heavy-handed with garlic, even to my taste, and I like it while Coley does not and has several times said so to Baliña, along with some other suggestions about the cooking.

"But"—I again—"but the señor has often said how good your cooking is. Only the other day in Comodoro I heard him tell some Englishman how lucky we are to have a good camp cook."

"Puede ser,"—the maddening, coldly doubting "perhaps" of the colloquial. "Puede ser, but here he is always complaining and telling me to cook differently. I know how to cook! I am a fine Argentine camp cook and I do the best Argentine camp cooking. The señor should never leave a hotel where he can get French meals. French cooking, indeed! Putrid meat covered with a sauce so that you cannot taste its foulness. I am a good cook! No one can tell me what to do! I am a wonderful cook! I am the best cook in the world!"

And so he carried on at length, his calm gone, his eyes glittering, brandishing the butcher knife with which he planned to kill Coley, who was still sleeping peacefully. Baliña became more and more irrational, abusive toward Coley and extravagant in his assertion of his own might and majesty and glory. The growing conviction that I was not faced by a passing, if extreme, attack of temperament but by an actual and deadly dangerous madman became a certainty. I argued with him.

"Stop, hombre, think what you are saying and doing! It does not make sense! You will come to evil if you carry on like this. Pull yourself together!"

"You think I am insane? Perhaps I am. But I know this, I will surely kill the señor! I don't know why I have put it off so long. Several times my head has turned and I have started to shoot him or to knife him. I can't stand it any longer! Probably I shall kill you both!"

But then suddenly his immediate fury seemed to leave him. He sat down and put his head in his hands.

"Perhaps you had better get another cook, Señor Doctor," he said more calmly.

"Yes, I think perhaps I should."

"But first I am going to kill the señor."

"Now look here, Baliña, old chap"—trying to be as fatherly as I could to a madman twice my age. "You aren't very well this morning. You're upset, and I'm sorry to say that you are talking without really taking thought. You must take time to lay your plans, to be sure that you know what you want to do and then to do it in proper style so as to be a credit to a man of your accomplishments, and not go off half-cocked and bungle everything. Today is the fourth. You think things over until the fifteenth. Then if you still want to kill the señor, I will make no objection and you can do it properly. Besides you can't kill a man before the fifteenth of the month. I dare say you never did such a thing in your life and never heard of any decent man doing it."

"No. Perhaps you are right, doctor. The fifteenth? Until the fifteenth then. I will wait."

"All right then. Now cook our breakfast and behave yourself. Not another word out of you about this until the fifteenth."

I went and awakened Coley myself, explained the situation sufficiently to put him on his guard, and we all had a quiet but somewhat tense breakfast together.

As I have already mentioned, the old ruffian has been getting moody and queer for some time, and I realized that he was not quite normal, but did not realize how serious it was and how near the breaking point he had come. He has had periods of sullen gloom in which he would mutter to himself, the only intelligible thing usually being something about not wanting to be a burden, that he did not count anyway, and who cared for poor old Baliña? Then he would have periods of gaiety when he was the soul of wit, when he was not "poor old Baliña" but Señor Ricardo Baliña, son of the grandees of Spain, the equal of any ten common men.

And to think that as he sat across the table from us in his somber mood, the old devil was seriously contemplating murdering one or both of us! Perhaps we never learn of our narrowest escapes from death.

The day of work at the barranca was something of an anticlimax after all this. Tonight Baliña is quiet. I think he will really keep until the fifteenth, although of course we will remain alert. The work here is nearly done and it would be too bad to leave it now until we are through, so I will plan on striking camp and moving as soon as we can, and on getting rid of Baliña then, which will be before the day now set for Coley's demise.

Feb. 5.

Stayed in camp all day, wrapping specimens, bandaging the reverse sides of several small blocks that had been brought in half bandaged, copying notes, repairing the car, and generally cleaning up some of the many odds and ends that remain when work at a successful camp is being wound up.

In the morning Paw Niemenhuis came and he stayed to lunch, the occasion being the return of the remainder of some medicine he had borrowed. What an unhappy family that is! Of course they are dead broke, but that is no peculiarity here and does not account for the atmosphere of doom that hangs over them. It may be partly their physical luck. Not long ago Paw and Maw were in a truck accident and neither has recovered fully; it was to massage his wife's badly lamed back that Paw had borrowed some ointment from me, although I warned him that it was in no sense a curative. The smallest girl has a broken arm which appears to have been set badly and will probably cripple her permanently. Another of the daughters probably has a cancer, about which nothing has been done, not even to the extent of a diagnosis. Another fell off a wagon and has been laid up off and on ever since. And then there is the third daughter, with a harelip and generally malformed. Only the small boy seems to be fairly sound, at the moment. Yet even this depressing list of misfortunes does not fully explain the feeling, which I have had about them from the first, that there is something malignant and hidden in the family. If I gave any credit to such psychic or supernatural factors, I might say that they lived under a curse—and so, in a sense, they do, but it is the curse of Patagonia and of the insufficiency for their own needs of their own abilities and temperaments. This is wrong with them, that they are not equal to the life that fate has given them, and perhaps—probably—there is something still more obscure and fundamental that is destroying them.

Toward the end of the afternoon, three Argentine friends from Comodoro arrived for a visit and they are staying the night. Coley's bed is made up in the back of the truck, I am sleeping on the floor of my tent, and the guests are established on the two cots and on the floor of Coley's tent.

Our three visitors are Irigoyen, Funes, and Lareta, a most heterogeneous trio. They went out with us to the barranca today and climbed down to look at our fossils. Irigoyen dashed about like a puppy, bringing worthless bits of bone to us. Funes looked the place over, grunted, and forthwith set himself to the task of climbing back up the adjective-adjectived barranca. Lareta sat down near me and engaged in long and philosophic discourse. These activities are typical of the three, who are as clear-cut types of Argentines as one could find.

Irigoyen is a pilot for the Aeroposta between Río Gallegos and Bahía Blanca, one of the most dangerous and difficult air routes on earth. He was the pilot whose plane hung over the Comodoro field, fighting the wind, for four hours the day before I arrived there. He is reputed to be one of the two best pilots in the country and has been flying for years, although still a young man. He is slight, almost thin, small, very dark, with a pinched, high-strung, very expressive, clean-shaven face. His energy is tremendous, but all nervous. He is never still, but must always be on the go, running about, mock-fighting with his companions, playing jokes. He is given to *aficiones*, whims or fads as violent as they are temporary. As soon as he saw our work, paleontology became an *afición*, and he swears he is going to spend all his spare time hunting fossils "now that I know all about it." Of course he will have a new fad next week. He had never spent a night in camp before, and to him it was a grand lark. How he envies us the opportunity to have such fun all the time!

Funes is a large and heavy man, indubitably entering middle age, but wearing well. The camp is no new thing to him. He is more the true *criollo*, the son of the country. He takes his maté amargo, bitter, while the others take it cocido, with sugar. He has spent years in Patagonia, and fancies himself as very *baqueano*, telling us at great length how to tie a horse to grass (loud gibes from Irigoyen!) and how to cook ostrich. He wears a shirt with wide stripes of yellow, pink, and purple, with his initials writ large across his ample heart. He smokes constantly and, alone of the three, drinks heavily. He is very blasé about being in the camp, and his interest in our work is perfunctorily polite. By profession he is a public auctioneer.

Irigoyen is the adventurer, Funes the bourgeois native and man of business, and the third, Lareta, is the aristocrat, the inactive, bureaucratic Portense (citizen of Buenos Aires) of good family and education, the thinker, the talker, and not the doer. He is humorless, but profound in a heavy way. In build he is like Irigoyen, but like an Irigoyen who has run down, flaccid and without energy either nervous or physical. Born of a Spanish father and an Argentine mother, his family could not afford to have him complete college, and of course it would not occur to him, or to any other Argentine of his class, to earn his own way through school. So he was found a subordinate post in a Bonaërense office, but the regular hours and the exigence of his superiors irked him so that he leapt at the opportunity for more independent employment, even though it meant exile in Patagonia. He cares nothing, he says, for business, but wishes to be a philosopher. Not, certainly, a philosopher of the modern type (as I conceive the term to include even myself) who seeks objective knowledge and then tries to interpret it subjectively, but one who would live apart from nature and reality and would seek knowledge only in his inner consciousness and in books by others of his ilk, expressing it in similar books and in conversation.

He is hopelessly out of place in the Patagonian camp. His sphere is the study and the salon. His attitude is defeatist. Patagonia, he says, is the last refuge of the failure, and he does not except himself from this condemnation. His stay in Patagonia is provisional, to be sure, but then everyone comes to Patagonia only provisionally, and most stay to die here. His fellow-countrymen (I continue to paraphrase his own ideas) are notable only for laziness and the only higher ideal is a post with no work and large pay at the expense of the taxpayer. The Spanish inheritance of the Argentine, which he personally shares even more fully than most, is tragic, for the Spanish character is not equal to the tasks of the modern world. He laments that he has parents and a fiancée, for what are these but iron bonds forever enslaving him?

Part of this is sincere but ridiculous self-depreciation, part is reasoned, and probably well founded, criticism, and part, no doubt, is insincere and merely angling for denial and for compliments. It was also his first night in the camp, for the Porteño ideal is to live in the largest possible city and not to leave it for any reason, and he took it seriously and with no pleasure. Paleontology has no personal interest for him ("Why don't you have peones to do all this work, and just stay in the museums and interpret the results?"), but he sees it uncomfortably as a small part of the interpretation of nature and the physical world which looms over his petty life.

The three of them, if my judgment is correct, are a remarkable epitome, almost unnaturally stylized, of the leading types of middle and upper class Argentines. In them one can see the basis for the greatness of the country, and for its weaknesses.

Feb. 8.

This afternoon we hauled all the heavy blocks out of the Oficina del Diablo, with the help of Duvenage and his partner Norwal. The truck was taken around by a long and difficult way to a place near the foot of the barranca, instead of at its top, and the men brought ten mules and two horses to help. From the Oficina proper, all the blocks but one could be taken out in paniers on the animals, and the largest block, too clumsy and too valuable to trust to the somewhat skittish beasts, was placed in a sling and carried down by hand, the way here being fairly easy.

The specimens in the second amphitheater, and especially the large block containing the snake, were much more difficult. To get the snake down the actual cliff, we roped it and lowered away, two men paying out while a third guided the block. Then we slung it so that all four could hold, and so carried it along a precipitous zanjón to a point where the most sure-footed horse could come. It was then loaded across the saddle and one man led the horse while two balanced the block, and so it arrived safely at the truck. Another block had to be treated the same way, but the rest could be carried by one man each. The mules were hitched to the truck, and with the help of the engine managed to pull it up out of the cañadón, and so to Duvenage's where we stored the blocks for the time being. On the way home we stopped at Van Wyk's puesto, near which we have a few specimens, found recently and now also ready to go.

All this field work is very hard on the hands. Ours are almost always somewhat bloody at night, and tonight they are like raw meat.

Baliña continues tractable, but far from agreeable or safe. However, the end of him is near, so far as we are concerned.

Comodoro
Feb. 9.

It took us most of the day to pack up all our goods and chattels, being particularly slow because Baliña was in a bad mood and the opposite of helpful. We stored the bulk of the camp equipment in Duvenage's sheep shed and here we loaded the truck with all the fossils brought up yesterday, bedding them on our folded tents for safety. We reached the Nollmann boliche at dark, had dinner there, and then decided to go on to Comodoro in spite of the late hour. At about the middle of the long climb up to the Pampa Castillo we got badly stuck, and it took us several hours to get out again. It was 2:30 A.M. when we got to Comodoro, and we had a hard time getting into the hotel.

Feb. 13.

The day after our arrival I paid off Baliña and sped him on his way, well in advance of the deadline, the fifteenth. I did not report him to the authorities, partly because that would mean as much trouble for me as for him, but chiefly be-

cause I do not see why he should not go and work out his own
destiny as best he can, now that we are safe from him. He
has a squaw and a hut up at the Laguna Pelada, and I saw him
started on his way there. That is wild country, even for
Patagonia, and there is plenty of room for him.

The problem of a new cook and camp man was solved,
after some searching, by the engagement of one Olegario
García Fanjul, a Spaniard, Asturian. He is twenty, Justino's
age, and is a friend of Justino's. He obviously has not
Justino's brains or ability, but he seems a good, useful sort
of lad and I hope he will do us. In any event he will not
turn out to be any worse than Baliña.

The rest of our time in town has been spent making up
accounts, buying supplies, writing letters, packing our fossils
in boxes and storing them at Bruzio's, having the car repaired,
and generally getting things in shape for a new start.

CHAPTER XII

PASO NIEMANN

Boliche Nollmann
Paso Niemann
Feb. 16.

WE were tied up in Comodoro, chiefly by the difficulty of getting replacements for the car, but left this morning. Our track out of Comodoro runs northwestward and climbs up to the Pampa Castillo in the long, oblique Cañadón Ferráis, near the middle of which is a boliche which has some importance because this is the main road north and is also a wool route from the interior. Here at the boliche we overtook a party consisting of Bishop Every, of Buenos Aires, his layman, named Dean, an English estanciero named Townsend, and an English schoolmaster from Trelew. Their car had broken down, so we stopped and Coley went to work on it, finding that it had to have a new cylinder-head gasket. The boliche was also in an uproar as someone was having a child, probably the proprietor's wife, although the proprietor seemed the only calm person around the place. Perhaps he is too used to it to be upset.

Coley, Townsend, and the proprietor went off down the Cañadón to get a gasket and a midwife, while the rest of us had lunch. By the time they returned the midwife was no longer needed, but the gasket still was and took some time to put on the car. We suggested that the Bishop christen the child, so that everything would be properly rounded off and we could go on our way rejoicing, but (from the parents' point of view) he is the wrong kind of bishop—Anglican.

This Anglican episcopate covers most of southern South

235

America, particularly Brazil and the Argentine. The Bishop comes down through Patagonia once a year, giving services in all the towns, borrowing any church or hall that there may be. In most or all places there is just this one service during the year for members of the Church of England, or the Episcopal church, although communicants are rather numerous throughout Patagonia. The great bulk of the population is, of course, Roman Catholic, as much as anything, but religion plays very little or no part in Patagonian life. Bishop Every is a very quiet and formal man, who confesses that our work might conceivably be interesting to someone, but not to him. Like many whom we meet, he cannot reconcile our field clothes and generally savage appearance with the thought that we might be gentlemen. After some time of conversation he hesitantly invited me to call on him in Buenos Aires, then added, rather too obviously seeking for some justification for this condescension, that he supposed that I did have a reputation. He was much taken aback when I replied "Yes, a bad one," to this peculiar remark. He is, however, a very good sort and I suppose we are a little hard to swallow in our present tough condition.

We finally arrived at the boliche at Paso Niemann, and retired early. A party of neighbors, from about fifty miles northwest of here, arrived for a *bochinche*, a blow-out, but we did not participate. Our room had two outside walls of the universal corrugated iron, and two thin board partitions against the kitchen and the common room. These were full of holes and with gaping cracks between the boards, so that the light streamed in like meteor tracks and constellations in our private sky. The visitors drank, danced, played the boliche phonograph, and otherwise and perhaps less innocently amused themselves until 7 A.M., this un-Patagonian behavior being due to our presence now in the Boer colony, which is neither Argentine nor European but combines the worst features of both. It did not help us to rest, as we were to all intents and purposes in the same room, the partition being more formal than effective.

Camp 4
Near Paso Niemann
Feb. 18.

Yesterday morning Nollmann and I scouted for a camp site while Olegario and Coley went up to Duvenage's for our stuff, and in the afternoon we located a place, unloaded the equipment there, dug out an old and poor spring, and then went back and spent the night at the boliche, this time fortunately without a bochinche. Today we set up the camp and scouted around our next point of attack, Cañadón Hondo, just west of us here, a broad and deep gorge emptying into the Río Chico a short distance upstream from Cañadón Vaca and on the opposite side of the Chico Valley.

From the river bottom, where the boliche is, the land rises in two abrupt steps. First there is a narrow lower terrace, not over a quarter-mile in width in most places, then a high terrace or low pampa, pampita, several miles wide, beyond which rises the great high pampa, the Pampa Castillo. We are camped in a rugged and sandy little zanjón, above the low terrace and cut into the edge of the pampita. Below us, in the zanjón bottom, is a spring, very muddy and not fit to drink. The three tents are on a narrow flat, Coley's and mine beside each other and the cook tent out in front, with a cache of supplies piled up between the three. On the side away from the dry watercourse is a wagon trail, not the main track and rather difficult to negotiate, and on the other side of it we have set up a service station, propping up a drum of gasoline on a mound of earth. It is a comfortable site, and the presence of the boliche within fifteen minutes or so by car makes us feel as if we were living in town.

We had a visitor for lunch, a real native type rare in this Boer region and making us think of the good old days at Colhué-Huapí. He wore the native rig, complete with black sash (faja), pantaloons buttoning at the ankles (bombachas), boleadoras, and a wine bag. The latter, called a bota, is a soft leather bag with a hard stopper in the neck through which runs a small hole closed with a screw-on horn cap or a wooden plug. To drink, the bag is held up in front of the face and is squeezed gently so that a fine stream of wine shoots into the mouth. This sounds simple, but in fact is a skilled operation, and an amateur is sure to get wine all over his face. Our visitor did not even need to open his mouth. He has a front tooth out, and merely lifts his lip and sends the wine sizzling in through the gap, a virtuoso performance which we watched with undisguised envy. In any case the bag must never be touched to the lips, and he lectured us on how sanitary and clean this is. I remarked that he doubtless drank maté amargo, using a single bombilla for everyone, and he admitted that this was so but went on undismayed to extol the cleanliness of the bota. These botas, incidentally, are always caked with tallow and very greasy, from being passed around while mutton is being eaten in the fingers.

In the good days, almost all the natives carried botas, but now few can afford even the common and very cheap wine. Except for this, our visitor's diet was quite criollo, mutton and maté, never anything else. He refused some delicacies as he said anything but mutton would surely make him ill after all these years. He is a very cheery soul, full of wise cracks, such as calling his horse "Camión" (auto truck) and his whip "Nafta" (gasoline).

Feb. 20.

In our reconnaissance of the cañadón we had seen and been beckoned by a green peak near the middle of the basin, unapproachable by car and some four or five miles from the nearest point on the rim, so today we left the car on that nearest point and walked to the peak. This cañadón is not, like most, elongated but is a deep, almost circular basin, emptying into the river valley below through a relatively narrow gap. The bottom of the basin, ten or twelve miles in diameter, is crossed by almost innumerable small, winding gullies, which make progress slow and uncertain. If one follows them, they soon prove to lead away from the objective, and striking out in a straight line means crossing countless miniature canyons, sliding down one cliff side and crawling painfully up the other. At intervals there are more open places with smooth, hard, sloping floors, ringed by hills of soft yellowish sandstone, carved into all sorts of fantastic shapes. Then perhaps will come a dark, narrow passage with perfectly vertical walls, suddenly opening onto a flat, floored with pebbles laid in sand, like a mosaic, varicolored and rich brown in general tone. Walking on these is not like walking on gravel, but rather like on deep velvet, or still more like the peculiar feeling of treading on mud when it is partially frozen but not yet quite hard.

Finally we arrived at a group of hills, almost small mountains, dominating which was our alluring objective, the great green peak. Such a place produces a weird impression, compounded of so many subtle emotions that it is beyond my power to reproduce or to describe it adequately. We climbed up to a pass between two lesser peaks; before us was a small valley, and beyond that the green mountain, completely bare of vegetation, its very steep slopes neatly laid out in horizontal layers. It is colored in pastel, here white, here creamy, but for the most part ranging from a true green to an opaque blue. One or two harder layers project as cornices, with vertical cliffs or shallow caves below. The top is capped by a flat bed of hard white stone.

We climbed up there and surveyed the landscape, then got down just below the crest on the lee side and had lunch. Later we went to a more outlying series of steep hills and crests, and here found some fossils, in part imbedded in the very hard silicified tuff, in part weathered out and lying on the slopes. The bones were neither very good nor plentiful, but sufficed to fix the age of the rocks, which is Casamayor. Coming back, we took a long swing around the other side of the cañadón, so that we walked altogether about five leagues, quite enough for one day over such difficult terrain. And so back to camp.

The most interesting creatures in this region are not the guanacos (still plentiful), the pichis, or any of the many that

I have mentioned before, but ants. This species is about a
quarter of an inch long, black, with unusually large heads.
They make nests in the sand and above them erect large piles
of small twigs, dead leaves, and the like. Whenever the wind
is not too severe in their sheltered areas, they may be seen
bringing up this material along definite routes. These are real
roads, two to six inches wide, readily visible because all the
small trash and larger grains of sand have been carefully
cleaned out of them. The stream of traffic is continuous, and
near the nest is so dense that traffic jams are common, there
being no rule of keeping to the right or to the left. The out-
going ants have no burdens and they thread their way through
the mob as best they can. The incoming ants are usually
carrying twigs, up to five inches in length, which they grasp
at one end with their powerful mandibles, laying the other
end so that it extends upward and backward. Others carry
small leaves, bits of grass, and even flowers.

When loaded they apparently cannot see well, for they
sometimes get swung around, badly obstructing traffic, or may
even take a few steps in the wrong direction, but they soon
correct this error. A puff of wind may blow the whole lot all
over the place, but in a moment they are back on the road as
if nothing had happened. On days of severe wind, transportation is suspended and the roads are quickly rebuilt in approximately the same places. A few ants forage at random,
but the majority go out to the end of the road system before
they strike out into a well defined foraging area in search of
a load. Sometimes a stick is unusually large and several
ants unite to drag it, but more often they work singly.

A typical nest, a few hundred yards from our present camp,
is the center of three main radiating roads. One is about
thirty feet long and the second about a hundred, these two
ending in small foraging areas. The third road is a great highway, six inches wide near the nest, over two hundred feet
long, and branching at the end so that it serves two large
foraging areas. This highway is, in comparison with the size
of the ant, about what a road ten miles long and forty yards

wide would be to a man. All the roads are as nearly straight as the ground allows, and seem to show more foresight and better engineering ability than any human road in this vicinity.

Feb. 23.

Today was spent prospecting new parts of the cañadón with Roux and Kock, two members of the Boer Colony here. At Roux's house there was a large gathering, his own abundant family, his father-in-law, and a visiting family of six. The place was simply bursting with children, all clean, plump, and seemingly happy. The children speak only Afrikaans, but the adults also speak English and Spanish. The Boers we have hitherto known were average for the colony. Some are worse, I am warned, and a few are better. This Roux family and its connections represent the aristocracy of the colony, and they rise above the criollo standard to about the extent that the general run of the colonists fall below it.

The house is of brick, made by the owner on the spot, and has wooden floors. There is a garden with tiny apple trees, each protected by a wind screen, unable to grow over three feet high, but each manfully bearing tremendous apples. There are also two windmills, and the grazing land is fenced. They are putting up a good, but hopeless, battle against the barren land and the ferocious climate. Their struggle is all the more remarkable, and more hopeless, in that the government has never actually given them title to the land. Their houses, windmills, fences, and all such improvements are "planta'o y clava'o"—planted and nailed down. At any time the land office can give them ninety days' notice and they must not only go, but also leave all the improvements, without recompense. When wool was high, the more provident families, like this one, put all their money into such improvements, thinking that changes in the land laws would enable them to acquire title. Now they are losing money each year, and their savings of former years are not their own property.

The whole colony, several hundred strong, came over from South Africa a few years after the Boer War. They are too clannish to make open accusation, but I gathered from hints and remarks among themselves that the migration was engineered by one or a few of their fellow countrymen who made specious promises, painted glowing pictures of certain prosperity, grew rich from the transaction, and then left the colony stranded here in a desperate situation, infinitely worse off than if they had never left home.

They all say that they abhor the native Argentines, and the chief hope of the elders is to get their children, all born here, away lest they intermarry with the criollos. In fact they plainly place the natives on the same plane as the niggers of their native land, and even refer to them as niggers. This ridiculous attitude is made still more repulsive (to me) by the fact that, in spite of exceptional families like the Rouxs, the average Boer is very definitely less agreeable, less reliable, and less fundamentally decent than the average criollo of these parts.

Old Kock, like others who were in the Boer War, is very proud of having fought the English and explains at great length that the Boers easily outfought their enemies man for man and were only overcome finally by the force of vastly superior wealth and man power. All still consider South Africa as their country, and they speak of it as if it had no relationship to Great Britain. Yet their hatred of the British is less than that they feel for the Argentines, and most of them cling to their legal status as British subjects and register their children as such in the consulate at Buenos Aires. They are completely discouraged with their lot here, and all live only in the hope of being able to return to South Africa. There has been some effort in their homeland to raise money to bring them back, but times are hard there, too, and the plan is not likely to be realized. The chances are that they are stuck here, and that their descendants will continue in exile, inbreeding and degenerating if they persist in their clannish attitude, until they come to bear the same resemblance to their ancestors as the half-witted Cajuns of the Louisiana marshes bear to the once proud and able Acadians.

Although they have never lived anywhere else, the children of the better Boer families are brought up to consider themselves as foreigners here. They are not allowed to learn Spanish, and what schooling they can get is from a peripatetic and third-rate Boer schoolmaster. Unless they do in fact succeed in escaping, this well meant course is certainly the worst thing that could possibly be done and the ultimate result for the children cannot be other than horrible. Some of the children, particularly of the lower type of family, rebel against this, learn Spanish, and are becoming acriollados in spite of their parents. The rebellion has no intelligent or even admirable motives, but I strongly believe that in the long run these children will be much better off.

Feb. 25.

We are working now in a low bank of green clay in the cañadón. This clay is very different from anything that we have found fossils in before, and the nature of the fossils shows, also, that it was formed under peculiar circumstances. At other places, for instance in all the diggings at Cañadón Vaca, at least ninety-nine percent of our fossils were land mammals, with relatively very scarce crocodiles and turtles and no birds. In the green clay there are only excessively rare scraps of land mammals, and we find abundant turtles, crocodiles, and birds. The inference is that this is the deposit of a lake or swamp where only flying or amphibious animals could come. Although they have become scattered and somewhat broken in the once soft clay, most of the bird bones are beautifully preserved. They are especially welcome, because they fill up a large gap in knowledge of life in the remote Eocene Epoch. No birds of comparable antiquity have hitherto been described from South America, even by Ameghino.

Feb. 26.

The wind has been terrific all night and all day today. Sand piles up in drifts in the tents and we can hardly eat or drink, everything is so full of grit. We find working in a sixty-mile gale quite normal now, but conditions on a day like this are prohibitive, even making walking almost impossible, so we stayed in camp, writing up notes and doing the other usual and everlasting small chores. In the evening we battled our way down to the boliche in the car for a short visit.

The bolichero, Guillermo Nollmann, is a small thin man, no longer young but always in a childish good humor. He was born in the Argentine of German parents. His wife is enormous. She is so fat that she looks as if she had been made of rubber and then pumped up almost to the bursting point. She was born in Denmark of German parents, early moved to a German island in the Baltic, and finally, I do not know why, came down to Patagonia.

Both understand German perfectly, but they speak Spanish, even when alone (an observation made possible by the thin walls of their guest room). The señora speaks it fluently but with a very strong German accent. She is so convinced that this is the correct way to speak that she becomes quite indignant with me because I refuse to copy her and say "garne" for "carne," "guando" for "cuanto," or "bichi" for "pichi."

Señora Nollmann closely resembles one of those little Stone Age goddesses that they dig up in the caves and rock shelters of Europe, the obese little statuettes that were apparently intended to symbolize female fertility. Yet she moves, a sterile mass in the midst of Patagonian fecundity. Like so many childless but maternal women, she takes an almost morbid interest in the many reproductive phenomena surrounding her. She raises pigs, fowl, and other creatures, and their sex lives are for her an inexhaustible source of interest and of conversation. She has a very foul little bitch, partly dachshund, and when we moved here Coley remarked that the animal was quite fat. "Oh," said the señora, with every chin quivering at the joke, *"that* isn't fat!" As the time draws near, tenderness fills her, and today she was making a bed for the purposes of puppy birth and assuring the animal that all will be comfortable and nice. If complete naturalness and real innocence can ever be obscene, then I have seldom seen anything more so than this gross but warm-hearted German woman imploring the rotund, nasty little bitch to come to her bosom.

At the same time, I have seldom known anyone more genuinely kind, unselfish, and whole-souled than Señora Noll-mann. In her capacity as godmother to her whole world, beasts and humans, she has assisted seventeen local children into the world. In an environment where men are usually selfish and often vile, she is genuine and a force for good, and I honor her as I laugh at her, or even when I occasionally turn from her in faint disgust. She and her husband are both kindly, pleasant, simple people. To us they are constantly helpful and neighborly.

March 1.

This camp is only about seventy-five kilometers from Comodoro, and today most of the people from Ocho came out on a visit—it is Sunday, incidentally. We gave them an asado, took them out to the bird clay to see our work, and to do some, and generally had an entertaining day.

March 2.

Piátnitzky arrived during the night, set up a small tent next to ours, and greeted us this morning. We all went over to the other side of the cañadón, and then Piátnitzky and Coley walked back, while I drove in state, as I am still weak from a slight fever for the last few days. It turned out that I had chosen the harder part, after all, for I got stuck and digging the car out single-handed was a long task. The others got back from their ramble after sunset, and were much braced because they had found mammals *below* the Casamayor beds, which means that they are the oldest fossil mammals ever found in South America.

Olegario was also excited, because our armadillo Florrie, who escaped several days ago, came wandering back into camp today. Incidentally, Florrie dotes on the big, black, road-building ants, chasing them and lapping them up eagerly with her long tongue.

March 5.

As I was having morning maté, a couple of guanacos on the pampita rim, black against the bright red sky, gave me an amusing shadow show. They would eat for a moment, then the neck of one silhouetted figure would rise and turn. The other looked up and yammered, "What do you see?" The first ably pantomimed, "I can't quite make out." Then both heads dropped for a moment, expressing "Well, I guess it's all right." Then both popped up together and the two beasts yammered at once, "Still, you never can tell!" And after a moment of hesitation their rumps swung around, "I think we'd better be going," and they bobbed off.

Also a very pleasant little bird has adopted our camp, accepting bits of fat and grease from us and repaying us with beautiful and varied song. Our almost inarticulate cook-naturalist, Olegario whose father is a García and whose mother is a Fanjul, tells me that this is a calendria, that calendrias like to live near houses ("Son muy caseros"), that they nest on the ground under small plants, that they live here all the year around, and that they are swell ("lindos") little birds. That exhausts his information, and it is a great deal for Olegario. Justino, on the other hand, would have gone on for an hour, describing the home life of the calendrias, their origin, history, and future prospects, their diet and color preferences, and their opinion of the protective tariff, in great detail—and more than half of it would have been true.

At noon two of the Botha boys came to call—I was in camp because I have been sick in bed for several days and still am (Piátnitzky has left and Coley is in Comodoro). The Niemenhuises and Duvenages are average Boers, the Rouxs and their kin superior, and these Bothas are the scum. There are several branches of the family, some worse than others, but none seem to be much good. These two boys never have been known to work, and they roam around sponging on more diligent relatives. They are quite elegantly dressed, complete with spiral puttees and silk neck scarfs, but they had no food, tobacco, or money and they asked me for all three, the first time anyone has begged from me here. Food is, of course, free to anyone who needs it in Patagonia, and we gave them all they could eat (rather a lot), and I even stretched a point and gave them some tobacco. They soon left, seeing that they could get nothing more.

The legend of the Great Clean Open Spaces, the moral influence of small towns, and the splendid characters developed by living out of doors was certainly invented by idealists who had never lived in a city of less than a million inhabitants. Even in our enlightened United States, I strongly suspect that the amount of sin and squalor per capita is considerably higher in the small towns than in the large cities and certainly criminal ignorance and dangerous prejudices are much more common in small communities than in large. People who live outdoors are usually decent at heart (so are people who live in big cities), but they are always dirtier and usually more petty than their more fortunate brethren who live in crowded apartment houses.—All this preamble to introduce the fact that here in Patagonia, the very apotheosis of bigness, cleanness, and openness, we have encountered probably the most loathsome and degraded family that I ever saw or heard of. I am charitable enough to omit the name, which happens not to be Botha.

This family lives near us here in a tent, which in itself places it low in the scale. Except during the summer and for nomads like ourselves, living permanently in a tent is really a very bad idea in Patagonia. Even they did not intend to; they got hold of this tent somewhere and set it up here to use while they were building a hut. Although they have done nothing else, after two years of work they have accumulated about a hundred adobe bricks, and now they are not making them as fast as the rain dissolves the old ones, so that there is not the slightest prospect of the hut ever being built. It takes about five thousand adobes

to build even a small puesto. In the meantime, eleven people are living in a single dilapidated tent smaller than our
cook tent.

All appear to be definitely subnormal mentally, and two
or three would probably not rate higher than idiots, scarcely
qualifying as human at all. All are broken out with loathsome
skin diseases, and several are permanently or semipermanently
invalided. The mother is the most thoroughly repulsive and
foul approximation of a human being that could be imagined,
so fat that I actually could not tell whether she was sitting
down or standing up, and with a gross, sagging, bloated,
purple-splotched face that has given me nightmares. The
most nearly normal-looking members of the family are the
youngest unmarried daughter, aged fifteen, and her three-
year-old son. This is not a full account of the depravity of
that family, but most of the rest could not be printed.

And members of this family refer to the criollos as niggers!

March 10.

Coley is back, I am up and around again, and we have
been back working at the bird clay for several days. Today
we spent the morning there, and then drove the car down a
circuitous old trail into the cañadón, trying to get nearer to
the place where Piátnitzky and Coley found the very ancient
mammals. We were soon completely stuck in sand dunes.
After two hours, we had the car out of that, but found that
we could not climb back up the grade, and so took our only
chance, that of driving the truck down a dry watercourse to
its outlet near the boliche. Almost immediately we were stuck
again, worse than ever, and at sundown we had made no
progress, so gave it up and walked to camp. There was no
moon, and we stumbled over the badlands and across the
pampita in inky darkness. The night was cold and clear, and
the beautiful stars shone brilliantly, and silently wheeled
through the black sky.

Summer is over. Five days ago we had a flurry of snow,
and even at noon now it is bitterly cold. The Río Chico has

ceased to flow, and water remains only in a few of the deeper holes in its bed.

March 11.

This morning we walked down to the boliche and told Nollmann of our predicament. He drove us off to a puesto, and then departed in his Ford in search of firewood. We went off with three boys who managed, after a rather poor exhibition with boleadoras, to collect five mules and their *yegua madrina*—the bell mare whom a troop of mules follows. With two riding horses and this troop hitched to a wagon, we went down into the cañadón to where our poor stricken car lay. The five mules and the yegua were hitched in pairs to the front of the car, a man on horseback led them, and another walked along beside the car, yapping and throwing stones to make the mules pull harder. Thus with sixty-one horse- and five mule-power, the truck reluctantly surged up from its deep hole in the sand and lurched tortoise-like up the side of the cañadón, a strange sight. All went well except for breaking our block and tackle rope, which was used for hitching up the mules.

All the animals were turned loose at the cañadón rim, and we drove the boys back home, stopped as short a time as politeness permits, and then went back to camp, reaching there at four in the afternoon. We had some lunch, and had just settled down for a siesta, being pretty tired by now and not having slept the night before, when Nollmann came walking into camp.

He was stuck, too! After leaving us he drove down the long abandoned road that used to skirt the Chico. He came to a very bad spot and stopped to throw dirt into a deep little gorge across the road. This done, he cranked his car, whereupon it started up, ran over him, and dashed into the gully, perversely picking out the deepest part. Like all Patagonian Fords, this one is in the last stages of disrepair, and the neutral does not work. While cranking, Nollmann has to wedge the thing into neutral with a block of wood, and in this case the vibration of the engine (no small disturbance!) jarred out the wood and started the car on its way, like a Frankenstein, to destroy its owner and itself. Happily Noll-mann suffered only contusions and some possibly cracked ribs, but the car was not only quite thoroughly stuck but also had broken both its front wheels and bent one at right angles to the other.

We dug and pulled the Fordacho out, with the help of our car, and took the most seriously bent or broken parts back to the boliche. The steering knuckle was heated in the kitchen stove and Coley forged it back into some sort of shape, using the zinc bar in the common room for an anvil. Eventually the bolichero's car was able to move very slowly under its own power, but with many things still wrong with it. It barely goes, and has so many essential parts loose, cracked or missing that it is a positive menace to its occupants, but he will doubt-less run it for another ten years.

The family has been born at last, one female and four male puppies, exactly like naked rats. The señora is in raptures.

March 15.

Our knapsacks bear the legend "Makes any burden a pleasure." The manufacturer would eat his words if he could have come with us a couple of days ago and helped us haul out our blocks from the bird clay, carrying them in the knapsacks for several miles to the car.

Now that that is done, we are concentrating for a few days on making a collection of the oldest mammals, which are excessively rare and very small, as extremely ancient beasts are wont to be, but all of which are new and worth their weight in gold. These occur chiefly on the slopes of a big white peak, which we call Mount Piátnitzky from its dis-coverer.

A mile or so away, also on the slopes of this peak, is a very strange place. Here the white sandstone has been eroded into pillars, hundreds if not thousands of them, all of nearly the same diameter. Most of them stand on a clean, gently sloping floor of harder rock. In some places, relatively low pillars, a foot or so high, look like headstones in a graveyard. Elsewhere they may be seen only partly weathered out of the surrounding sandstone, and they stand like half-columns against a wall. In another place dozens of them rise freely to heights of five or six feet, and they look like the columns of some half-ruined Egyptian temple. The whole thing resembles the relics of an ancient civilization, and it is hard to believe (although necessary to believe) that it is wholly due to the forces of nature.

Before dinner we stopped at the boliche for a few minutes to square up our accounts, as we will soon be leaving, and to gossip. They had a big asado yesterday, mostly people out from town, and we joyously inherited some leftovers, including a chicken, some beef, some fresh vegetables, and other rare and imported delicacies. The old criollo who cooked the asado was still there, and there also dropped in a lad whose only claim to fame is a recent attempt to kill his father.

March 19.

Our work from this camp has entered another, and final phase. Piátnitzky had pointed out to us a place where he once found some fossil fish, and we are making a quick collection from there before leaving. This rounds out our collection nicely. From the Casamayor, our especial field of attack, we now have large collections of mammals, reptiles, birds, and fish, as well as smaller numbers of fossil plants and of fossil shells. This will give us a knowledge of the life of that time almost unequaled in its scope.

This morning we drove up to Kock's puesto, on the far side of the cañadón and nearer our fishing grounds, and thence worked our way down into the cañadón by car and on foot. Fishing here in this waterless desert consists of digging out thin plates of limy shale and breaking them up carefully by splitting along the bedding planes. Every twentieth or thirtieth split brings to light the paper-thin impression of the skeleton of a small fish, a few inches long. After a day of this strange sport, we drove back to Kock's and found that he had prepared a very good asado for us in back of the house. We sat on the woodpile eating sizzling roast mutton, smoking, talking, and generally having a good Patagonian time, and then set up our beds in the living room and retired. Kock has two passable young daughters, but they are not here tonight—perhaps his quaint definition of niggers includes us and he has sent them away for safety.

March 21.

Most of this day has been spent packing, and tomorrow we leave Camp Four. We will set up no more base camps, but from now on will travel light and spend only a few days, at most, at any one place.

This evening we paid our last visit to the boliche, for a time at least. As usual they had a visitor, and as usual an interesting one, a little mustachioed man as dark as an Indian who had been passing on horseback (living "allá arriba," which means anywhere between here and Buenos Aires), and who dropped in for some sardines on the half-tin and a carafe of Mendoza wine. Señora Nollmann had an ache in her barrel, as she puts it with typical innocent indelicacy, but she says she cured it by staying up till two o'clock last night drinking brandy. Now the ache has migrated from her barrel to her head.

The wind was blowing, of course, and whenever the conversation lagged the señora would shake her head and say, "This wind is going to bring rain!" That is her invariable weather prediction, day after day, and sooner or later she is bound to be right. Olegario, an incurable optimist, is also a consistent weather prophet, but he always predicts fine weather, and as a result his batting average is poorer, about .001. Each morning he is sure that this is just the one day when the wind is going to die down. As for rain, he simply scorns the idea and continues to say "These are just fair-weather clouds" until fifteen minutes after the heavens have opened and we are drenched to the skin. Then he grudgingly admits that it is perhaps going to sprinkle a little after all.

Among our many effects are now three active, reasonably tame, and very healthy pichis. The inveterate and favorite Florrie is still with us, and Nollmann has at different times given us two more, larger ones. He intended them for our larder, but we found that they would follow Florrie's example and eat, so we are keeping them as pets. Olegario is much taken with them and usually has them running around the cook tent in their ridiculous, prehistoric way.

CHAPTER XIII

NORTH OF GOLFO SAN JORGE

March 23.

YESTERDAY afternoon we drove up the long and now familiar ascent to the top of the Pampa Castillo, and then, instead of turning right and down the road to Comodoro, turned left and continued on top of the pampa along the picada to the grandiosely named Hotel de La Pampa de Eduardo Cuestas. The picada is the main, one might say the only, road of Patagonia, in contrast to the many self-made tracks and trails, *huellas*. From Comodoro it climbs to the top of the pampa up Cañadón Ferráis, and then continues north, eventually to Buenos Aires. It has much traffic, for Patagonia, that is, perhaps a dozen cars a day near the towns, and in places it has been graded. The sad fact is, however, that the grading does more harm than good. On top of the pampa the cross-country going is not bad and the grading has simply removed the top layer of pebbles and exposed the underlying clay, with the result that wet weather makes the picada completely impassable. When it has been raining, everyone perforce abandons the graded road and drives alongside it on the virgin pampa.

This morning we set out to find Cabeza Blanca, a famous fossil locality, and took one of the local lads, the usual swarthy ruffian, as a guide, with poor Olegario tossing about in the back of the car. Our guide finally confessed that he had never been to Cabeza Blanca, but explained that that was not his fault as he had never even heard of the place. At length we came to a puesto in one of the cañadones running down toward the Río Chico. Here three men were dipping sheep, while an asado sizzled over a small fire, and we had a convivial maté with them and received better instructions regarding our destination. About one kilometer from there, the truck developed bad symptoms, so we stopped and lunched on martineta and ostrich and then the local lad and I went over to the hill while Coley and Olegario worked on the car. I located a few good fossils, shellacking them and leaving them, and by sundown the car was fixed and we returned to the boliche—pardon me!—the hotel by a different and longer route.

March 26.

The next day, the 24th, we set out for several days in and around Cabeza Blanca, and came back to the hotel tonight. We went first to the Boer puesto nearest to our fossil locality and there purchased a sheep. At Cabeza Blanca, Coley and Olegario put up one tent while I took out the things I found yesterday. In the afternoon, after an asado, we all turned to and found a few more good specimens.

As this was such a short stop, we crammed three cots into one tent, and for a kitchen Olegario threw up a small brush windbreak. Seated on sheepskins, we ate dinner there in the open the first night while an unusually blatant sunset raged overhead.

Cabeza Blanca was, I believe, the first place in Patagonia that I knew by name, and it certainly is one of the most famous, being known by scientists all over the world. Yet the local inhabitants do not know it, and it was not until we located the man who actually owns the land that we found anyone who had ever heard the name. This is not really surprising, however. In a country so full of great peaks and cañadones, Cabeza Blanca is a very minor physical feature: an isolated hill about two hundred and fifty feet high and about half a mile long. Its fame in foreign parts is due entirely to the chance that it is, or was, an amazingly rich fossil locality, a point which has no interest for the natives. It was discovered by Carlos Ameghino, and later the American, Loomis, spent some time here and made a large collection, now at Amherst College.

The number of fossils that have been found in this small hill is astounding, but we do not propose to settle down and work long here. In the first place, the fossils are all of Deseado or *Pyrotherium* age, hence much younger than the ones we came to study particularly, and they are already so well known that we could not hope to contribute much. Then, too, the collecting is not now as good as it was. The deposit is inexhaustible, for all practical purposes, but it is necessary to allow time for a new crop to develop. After being intensively worked, it would have to be left for ten or fifteen years so that a new crop of fossils might weather out. Unfortunately, however, it is so famous that every traveler to these regions who has the slightest interest in geology goes there and picks up what he can, with the result that everyone gets a little, but no one since Loomis has had a chance to make a really large collection here.

We finished up what we wished to do at Cabeza Blanca itself yesterday, and today we went a couple of leagues down the valley to another hill. Although less fossiliferous, this is much more impressive, rising in abrupt steps about four hundred feet above the pampita. There are long clay slopes, perfectly vertical cliffs of ash, deep pits like craters, twisting zanjones with natural bridges and long underground passages, all on a herculean scale. One often hears of lunar landscapes, but this is one of the few that might really be on the moon. Certainly it is unearthly.

We slid and climbed around this all day, and then came back to the Hotel de La Pampa, arriving just after sunset, which was gorgeous. Today was even windier than usual, and it is the rule here, as if in compensation, that the windier the day the more brilliant the sunset.

Perhaps this place could be called Camp 5, being our present base, but even though it does not deserve its ambitious designation of "hotel," it hardly rates as a camp. It is even superior to any boliche we have seen, and in local hierarchy should probably be classed as a fonda. It is built of brick, a few burnt and the rest sun-dried, and the central part is very unusual in possessing a slanting roof. There is, of course, a bar and common room; then there is a cellar and some storerooms, dining room, kitchen, owner's bedroom, and three guest rooms. Of these, one has four beds and is for hoi polloi, another has two beds and is for the élite (to wit, at present, us), and the third is now full of our camp equipment and odds and ends, together with a few very dejected vegetables piled in a corner.

Around the place there are several trees, one almost as high as a man. All water has to be hauled several miles in a barrel mounted on two Ford wheels, pulled by a sad little mule and driven by a bandit named Palacio who also serves maté, waits on table, says "Buen día" to newcomers, does the gardening, tends the animals, and repairs motor cars, all without the aid of a net. The inn was formerly run by an Italian named Zanotti, but less than a year ago he sold out to one Eduardo Cuestas, a superior Argentine of Spanish ancestry, young and hopeful. His handsome and very pleasant wife is a tiny, extremely brunette Chilean.

Señora Cuestas has just finished making me a new tobacco pouch. When we first went down to Cabeza Blanca, Coley shot an avestruz which he presented to Trujillo, the able guide who got so badly lost that we had to find his way back for him. Trujillo neatly skinned the ostrich's neck, taking off the hide in a tubular piece like a sleeve, and presented it to me. Olegario cured it, a simple process of scraping off all the grease, drying thoroughly, and then softening by rubbing and folding. Señora Cuestas then tailored it and stitched it up for me. Sewing straight across the large end produces a pocket and then the rest is cut into two long streamers which are wrapped around the bag to hold it shut. The leather is very much like old, grey, translucent parchment, except for the pattern of small holes where feathers were.

Now, although it is after midnight, we are sitting up because Cuesta insists that he is going to get us North America on his radio, a machine which in itself stamps him as a fairly prosperous newcomer in the country. So far it has produced nothing but static.—There it is! From the crackling and wailing there emerge a few scraps of words in English, part of a hymn, other fragments of music, then a chime and a very faint but intelligible voice saying "This is Station W——, Chicago." So Chicago really does exist still! Outside the wind is howling and clouds are scudding across the dark Patagonian sky, hiding the Southern Cross. On the monotonous Pampa Castillo nothing can be seen. It is many leagues to the next habitation.

So to bed.

March 28.

Coley had not liked his bed much and so had another set up for him somewhere else and left me alone in our bridal suite. The door was banging in the wind, so I locked it tightly and was just dropping off to sleep when someone began pounding on it. At first my visitor refused to say who he was, and finally said that he had come to sleep in the bed. "Well, who are you?" "It's I!" "And who is 'I'?" "The person who has come to sleep in the bed." This bright conversation continued for some time until I consigned the unseen to the devil, rolled over, and went to sleep. This morning it proved that it had been Cuestas' young brother and that he had traded beds with Coley, without my being informed of the deal. The poor lad had to sleep on a sack of potatoes in the storeroom. Too bad, but if people want to come into my room they will have to learn to give a coherent account of themselves.

Today we went down to a place called Lomas Blancas, White Hills, a few miles below here on the eastern, coastal, side of the pampa. Both topographically and geologically it is one of the most confusing places imaginable, with zanjones running in every direction and the rock strata tilted and slid about until they lie at all angles. After lunch, Olegario and I climbed the highest point to get an idea of the layout, while Coley, as I thought, skirted its base looking for fossils. We worked around a little, not finding anything of great interest, then had a look for Coley but did not see him. We returned to the car, confident that he would be there since it was time to leave if we were to get out of the badlands by dark, but still no Coley. We drank some maté, and by that time it lacked only an hour of being pitch-dark. I left Olegario near the camión with instructions to light a fire after dark and to shoot into the air if he heard a call or signal, and I went off to where Coley was last seen and scoured the surrounding badlands. By this time I was very much worried. The place is so confusing that the chances of getting lost are great, and this might or might not be serious. The possibility of severe injury in this region of treacherous cliffs is also a constant menace, and an injured man would be so hard to find that he might well die before aid could reach him. Much less likely, but very definitely possible, was a chance of foul play. During the afternoon one of the surly homeless men had wandered past on horseback, and only last month two men from Comodoro were found shot to death in the badlands, presumably the victims of one of these wanderers.

When it was too dark to see, I started back to the car. Just before reaching it there were three shots, and when I came up there was Coley, safe and sound, shooting to announce his arrival and recall me. The least serious of the possibilities had occurred: he had gotten turned around in the broken country, had wandered awhile, discovered that he was thoroughly lost, so climbed up to the high pampa, a tremendous climb, seen the car from afar, and finally managed to make his way to it.

March 29.

I have neglected to mention the house mouse and its tame cat. These two were great friends and played together by the hour. The cat would chase the mouse, which made no serious effort to get away, and when it caught the mouse would carry it about in its mouth awhile, then release it and begin the chase again. The mouse apparently enjoyed this just as much as did the cat. Several times when the cat had tired of the game and refused to play any more, I saw the mouse come up and stand directly in front of its friend, obviously trying to entice it into continuing the game.

Now this idyllic friendship has come to a tragic end, as I suppose such ill-matched liaisons always must. Coley decided to join in the fun and started to take the mouse out of the cat's mouth. The cat was very much annoyed at this intrusion on its prerogatives, and emphasized its displeasure by holding on a little too hard, with the result that the mouse was killed. The poor cat will now be very lonely, as I suspect that it will be a long time before it can find another mouse so trusting.

As we started out this morning, a grey little man in store clothes begged a ride from us. He had been coming up from the coast yesterday in a Ford which refused to function. His companion made a fire of dung and spent the night drinking maté, while the little man had been walking and climbing all night trying to find help. We finally reached his car, if it can be dignified by that name; the most disreputable old Fordacho in this graveyard of the living dead among Mr. Ford's products. It absolutely would not go, so we towed it some miles to the nearest house and left it there in the corral, with its owner sitting on the running board with his head in his hands.

We went on down the Cañadón de la Máquina, so called because there used to be a sheep-shearing machine there, and came out on the slightly rolling coastal terrace close to Puerto Visser. In other days, before Comodoro became the metropolis of Patagonia, Puerto Visser was an important town, with large stores, inns, bars, warehouses, lights, laughter and gaiety. Now it is an indescribably sorry sight. Most of the buildings have disappeared entirely or are represented only by mouldy foundations. Only two are still occupied, one of which, as we drove in, showed no signs of life while the other was overflowing with what appeared to be about twenty-five children all of the same age (a good trick, whoever did it). Then there is a large corrugated iron warehouse still standing, but completely empty.

The only other occupied building is a police station, the Lord only knows why. A very ugly Indian nattily attired in a dirty wool undershirt and a policeman's cap made for a much larger head rushed out like a gouty snail and dully watched us pass. The comisario himself, looking as if he had crawled out from under a stone where he had been hibernating, poked his long and dirty nose out another door, but hurriedly withdrew it again. We stopped above the shore cliff for a few moments, watching the gulls and a couple of the snappy black and white porpoises so common here. There is no harbor at all, only the open sea; but of course that is true even of Comodoro. This ruined town, once a busy port and now with only the last traces of senile life, is one of the many signs that Patagonia is tending to retrogress rather than to advance. Even in Comodoro, incomparably the most active place, everyone complains that things were much better ten years ago than they are now.

Near this corpse of a town, there is an enormous, almost perfectly round hill, its sides beautifully displaying the geological strata like layers in a cake. We studied this quite thoroughly, spending the whole day at it, and made some observations of great interest. At least one was sensational, for we found a few scraps of mammal teeth in beds considerably older even than those in Cañadón Hondo, and hence the oldest mammals yet discovered in South America. Coley, who has been under the weather a little, seems to have lost his sense of direction temporarily, for he got lost again. Fortunately I spotted him with the glasses, walking in exactly the opposite direction from what was intended, and we quickly caught him. Considering the fact that we often separate in the badlands and that it is so extremely difficult to keep oriented in them, we have been extraordinarily fortunate that no one has been seriously lost so far.

April 1.

We have polished off the cliffs north of Puerto Visser, all that remained to do within striking distance of the Hotel de La Pampa, and today we left. Last night there was quite a bochinche, although not in celebration of our departure. It started with a brunette gentleman who had run up a large bill there and left some clothes in security. He came back to get the clothes, which he visibly needed badly, but without the means of redeeming them. Cuestas could not see his point of view, and the debtor worked himself up into a purple passion, using some splendid words which I have duly noted against future need. He finally left, breathing brimstone and muttering promises to come back with reinforcements mañana. Then several trucks, engaged in hauling along the picada, stopped for the night and their drivers played cards until a late hour. The proprietor's brother, who is thin to emaciation and hence is called Gordito ("Fatty"), had to sleep on the floor in the dining room. This is his vacation from school, but he is not enjoying it very much.

This morning while we were loading up, a car arrived with prisoners and some police from Sarmiento, and I had a chat with a prisoner named Williams, brother of the godfather of Casa Williams at the source of the Chico. He and several others are under arrest because of a complaint that the local Comisión de Fomentos had misappropriated funds. The governor ordered all of them to Rawson, the territorial capital, for trial. Williams just happened, says he, to be far away in Esquiel when the police came for him, but they got him finally. The car in which they are driving is his, and he has to pay all the expenses of the trip, including those of his armed guard. Of course he says that they are all innocent. Perhaps they are, at that.

Then a man came riding up on a horse which he loudly claimed to be the fastest in Patagonia. Cuestas bet him two bottles of cider that one of the local horses was faster, and a race was arranged forthwith. From where I was standing, the stranger's horse seemed to win by a length, but the local lads raised a dispute too technical for me to follow and the bet was not paid.

After this stirring event, we took the road north, along the picada, where we passed two cars badly stuck and one moving slowly, and eventually arrived at Malaspina. For the first time in my experience, local informants overestimated a distance. They said it was fourteen leagues and it was only twelve and a half.

In Malaspina we were delighted to see a large sign proclaiming that here was a hotel de luxe, with the finest accommodations and food to be had south of Buenos Aires, but when we reached this elegant place (not quite as elegant as its sign) there was no evidence of life. Finally after long beating on the door, a very cross-eyed man stuck his head out a window and told us that the hotel is closed because the proprietor is in jail. For murdering one of his guests, probably. This reduced the choice of stopping places to the vanishing point, and we went necessarily to the Hotel de los Vascos, in the guest book of which I wrote (in English), and with more charity than might be imagined, "This is the most cheerless, God-forsaken, barn-like, clammy dump it has ever been my misfortune to find myself in." The boliche at the Confluencia of the Mayo and the Senguerr, which I have hitherto rated as touching bottom, at least had no pretensions, whereas this really looks as if it might be something. It is the sort of place in which you are always looking over your shoulder to see whether the crime has occurred yet.

The kindest description of Malaspina is to say that it is not so bad as Puerto Visser. It has a telegraph station, on Patagonia's single line which follows the picada, one store (closed), a nasty boliche that calls itself a hotel, and two or three houses. I do not see why they want a town here, anyway. Its one grace is the presence of lots of water, seepages from the pampa rim, and consequently of a big weeping willow tree.

The police and prisoners arrived long after dark, having been stuck several times on the way.

April 3.

One night at the Hotel de los Vascos was more than enough, even aside from the fact that we found our beds already so fully occupied by vermin as to leave little room for us, so we abandoned Malaspina yesterday and worked our way down to the Estancia Las Violetas, which is, anyway, nearer the things we came to see. The barrancas here, forming the east slope of the pampa alta, are very grand, vertical or nearly so in most places, and forming an almost continuous, very sinuous line, more like a single long cliff and less like a strip of badlands than most of the barrancas of the region. In the time available yesterday, I began taking geological profiles and Olegario, to the surprise of everyone, found a very fine fossil crocodile skeleton.

Last night they gave us some puchero at the estancia and a room to sleep in. This is the first large estancia that we have seen, and it forms a pleasant contrast with the little puestos. It belongs to Alfonso Menéndez Behety, son of the Menéndez who founded the Anónima and hence a member of the family which owns so much of what is worth owning in Patagonia. He is, of course, an absentee landlord, although a large room is kept sacred to him. The estancia includes about 140,000 acres of camp, some of it way above the average for grazing. The buildings are extensive, including shearing sheds, storehouses, stables, workers' barracks, cook house, and the administrator's house. We were, I confess, rather piqued that we were not allowed to sleep in the latter or to eat with the gentry, but then we keep forgetting that down here anyone who dresses roughly and does any physical work, both of which we do to excess, is classed on sight as a peon. This morning the administrator relented a little and took us through the house, an enormous two-story concrete building, unfortunately full of large cracks and generally rather moth-eaten. We were even allowed to gaze, for a brief moment, into the holy of holies, the owner's room, my quick impression of which was of a great, dark cubicle full of faded red plush and grimy oil portraits.

The administrator is a Scotchman, apparently very keen but overendowed with the dourness supposed (and usually incorrectly, in my experience) to be characteristic of his race. His wife is a queer old soul who never quite understands and who fills in pauses by saying "Sí, sí, sí, sí, sí, sí," or "Yes, yes, yes, yes, yes, yes," very rapidly and apropos of nothing. I told her that we had found a fossil crocodile in the barranca below the house, and she said "Oh, sí, sí, sí, a crocodile," as if she had to shoo them out of her kitchen every morning.

"Yes," I said, "a very ancient one too. It is over forty million years old"—a great understatement, but I wanted to let her down easily.

"And is it dead?"

(I know Mark Twain had a similar story long since, but this is perfectly true.)

Hence we went down to the barranca to collect our crocodile, and I think we must have come close to hanging up a record for collecting a specimen of such size, about six hours working time, all in one day. In the morning Coley and I got the whole thing blocked out and shellacked. In the afternoon, with Olegario's assistance, Coley bandaged the top, dried it by building fires around it, turned the block, shellacked and bandaged the under side, carried it to the truck using the tail-board as a stretcher, and there dried the under side by a fire, so that it was quite dry and ready by the time I got back.

In the afternoon I set out to study the barranca, which I did in the course of a twelve-mile walk, climbing up and down the slopes, collecting fossil shells, and so on. I climbed up to the top after a while, and walked along it for some distance. When it came time to leave, I looked around for an hour trying to find a way down the cliff and could not, so I finally became angry and went down anyway, which I record with shame as it was one of the silliest and most foolhardy things I ever did. I have never had such a bad climb, and it is only by the grace of God that I am here tonight. In one place, for instance, the only way was to jump vertically ten feet, landing on a slope smooth and much too steep to stand on, slide down that about twenty-five feet more, and then stop, if at all, on a ledge a few inches wide above a direct drop of at least fifty feet. If I had hit the slide too far out, the thin sharp edge of crumbly tuff would probably have given way and dropped me vertically altogether about one hundred feet. I am not the intrepid type, and as I went down I was scared stiff, more frightened than I have been since the revolution, but all went well and the only injury was to Coley's field glasses, or rather to their case, which suffered severe abrasions. From the ledge I could worm my way to the top of a practicable crevice, and so to the bottom, and thence weary miles back to the car.

Shortly after starting, we got stuck, a routine matter, but in this case it used up all the time we had intended to spend to get to Bustamante, our next destination, before nightfall. Travel here in uncharted country is very inadvisable after dark,—on our way this evening we passed an abandoned car standing on its head, thoroughly wrecked, in a deep gully right across the track. To make matters still more cheery, our head-lights burned out almost immediately. We tried to keep on, feeling our way along and sending ahead a scout when things looked bad, but of course the inevitable happened and we were soon stuck, although in a very nice way, merely having the wheels fall into deep ruts so that the car was supported on the axle and differential. We decided that was enough, and here we are spending the night.

It is an old Patagonian custom. Everywhere along the roads there are little clearings, each with a pile of ashes. Here a car has broken down at night and the driver has had to stay out—trasnocheando, they call it.

As a matter of fact, and as usual, we really are in very good luck. The car is not damaged, nor are we. We have food, a cook, and beds with us and have just had a very satis-fying dinner, al fresco. Our cots are set up in the road, danger of traffic being not very serious, and we will soon be asleep. The night is wonderfully fine. The wind has moderated and there is a splendid moon. All around us in the darkness the martinetas are whistling to each other, like frogs in a pond.

April 4.

The car was very easily put back on its wheels this morning and we went on our way rejoicing. Our first stop was at the Tetas de Pinedo. [Preparing a lecture once in Buenos Aires a refined friend urged me to call them the "Mamelones," that being a more elegant word, but tetas they are to the local people, tetas they are on the official maps, and so tetas they shall be in my work.] These are two large rounded hills, standing near each other and rising above the coastal plain with an appearance, as the name implies, extraordinarily like two gargantuan breasts. That they are composed of particularly hard rock is a fitting allegory of this infertile land.

We climbed the east teta and were rewarded by a magnificent view, on one side the cliffed border of the northern end of the Pampa de Castillo, and on the other a series of terraces stretching away to the coast, which here makes a broad sweep eastward, the northern margin of the Golfo de San Jorge.

Hence we tried for Bustamante again, and this time succeeded. Here, too, there is no reason to worry over choice of hotels, since one Vicente Morant has a complete monopoly. His boliche is no better nor worse than the average, but it is below par as to stock and accessories. For instance, there was just one bottle of beer (and that flat) in the entire place and the table wine is without question the worst that I have encountered in ten countries or on the high seas. Don Vicente and his lady are very dejected souls who look as if they expected the worst, an expectation not far from fulfillment.

This afternoon we ran out on the Peninsula Gravina to its tip, Punta Ulloa, much enjoying the salty air and the seashore after our long stay inland. The rocks are brightly colored porphyries, much older than the beds in which we have been seeking fossils. The shore is strewn with the bones of stranded whalebone whales.

We now have three pichis. The one caught last January,

Florrie, is still alive and well. Our others died at Cuestas' boliche, but since then we have caught two more, one today. Florrie is a great social success wherever she goes, and here she has the run of the house.

Solano
April 8.

After leaving Bustamante three days ago we worked our way down along the coast to Puerto Visser, then back up to Cuestas' place, and hence finally to Comodoro, which we left again this afternoon. Earlier in the day we went out with Luro, one of the Aeroposta pilots, to see the plane come in from Deseado. One of Luro's friends came out at his house here in town to greet us, dressed only in pajamas, and Luro dragged him into the car and carried him out to the field, to his intense distress. He spent most of his time trying to be inconspicuous in the back of the car.

The plane was landed successfully in spite of the bad wind. The soldiers got ropes hooked onto it the second it touched ground and managed to pull it into the hangar before it could blow away. I was impressed as never before with the strain of piloting a plane on this run. When Irigoyen got out I did not immediately recognize him. He looked seventy years old. His face was haggard and his hands trembling, and at first he seemed to be in a daze and hardly conscious of his surroundings in his reaction to the exigencies of the flight, which had been a terrible one, with extreme wind and the motor working badly. Soon, however, he recovered his customary high spirits.

There was one passenger, an estanciero named Rudd from southern Patagonia. He knew Hatcher and Brown when they were collecting bones in the nineties, and he is quite au courant of the fossil business. We had lunch with him, Luro, Irigoyen, and the local manager of the Aeroposta, and shortly after drove out here to Solano on the strength of an invitation from Deckers to make that our headquarters when we came to study this part of the coast. Deckers was not here, but we moved in anyway, Olegario cooked us some dinner, and soon after our host turned up and told us to make ourselves at home, a welcome which we had somewhat anticipated.

April 12.

Three days have been spent for the most part on and around Pico Salamanca, a high conical peak rising right out of the water, which has been a famous landmark since the days of Magellan. Such distinctive objects are few on this rather monotonous coast and many an ancient navigator had reason to bless Pico Salamanca as he tried to beat his way along this rocky, harborless shore with its tremendous gales, thirty-foot tides, and treacherous currents. It was less generous to us, yielding no fossils of any value, although some other information of considerable importance in our work.

This morning we went over to a wildcat well which the government is drilling along the north scarp of the Solano basin. Work was suspended, and we suddenly realized that today is Sunday. The only person in sight was the chief of this operation, who appeared at the door of his tin hut, clad in pajamas, so we went on about our business. Very soon a much upset lad came and told us to report to the chief at once, so back we went to see him, his pajamas, and all. He requested us firmly to leave the place and never come back, but I told him that we are the favorite children of Mr. and Mrs. Yacimientos Petroliferos Fiscales, and we went back to work. In about an hour we passed his hut again, finding him still in pajamas but now very affable. His mate suddenly remembered that he had seen me looking at well samples (which are holy and secret) with the big shots at Tres, and we were told that the place was ours. Apparently what really happened was that the chief called Tres for instructions or for reinforcements to help throw us out, and they told him that we were crazy but harmless.

Working near the well we soon excited much comment and attention from the employees. I had just found a fossil level and was running through a lot of fragments to pick out any teeth that might be present, when up dashed two peones, panting hard.

"Have you found it yet?"

"Found what?"

"Gold, of course."

Even when I showed them what I was picking up they remained very suspicious and wanted to know whether fossils were sold by weight or by the piece. The peon class here is obsessed by the idea that there is gold about, and their idea of a gold mine is that the discoverer just picks up the gold, puts it in his pocket, and walks off quietly. Any one of them will tell you that enormous quantities of gold have been found around Comodoro but that the strikes have all been kept secret, and of course they all think that is what we are doing. In vain I have assured them that I am an Ingeniero de Minas (a lie), that I know all about gold (another), and that the chances of finding gold in paying quantities here are less than one in a million (the truth). They "know," with that curious popular ability to translate hope into "knowledge" that has done so much mischief in the world, and nothing can shake their conviction that they live in the midst of riches which are being secretly filched from them.

Comodoro
April 13.

Our work in Chubut, along the northern half of the Golfo de San Jorge, is now done, and as soon as our fossils are all stored safely, car fixed, and other chores attended to we will start a brief excursion southward, into the Territory of Santa Cruz.

Here at the Colón there is the usual agglomeration of unusual people, for instance one Cosandier. He is a young Swiss who was an expert draftsman in Geneva but suddenly decided that he would like to raise sheep in Patagonia. With one companion he arrived here just the week before I did and he set out for Colonia Sarmiento in an old car. The unhappy lads spent ten days on the roads, although the trip can be made in a day with good luck, and I now learn that they never did reach their destination. After seeing a little of the country and of the misery in which the poor puesteros live, Cosandier's vision of a quick fortune in sheep faded. He is now trying to get some sort of job in Comodoro to see him through the winter, and next year he is going back to Geneva unless things have picked up here by then. So he says, and I hope he does it, but it sounds like the typical beginning of a lifetime of exile. With few exceptions, settlers who do not leave at once stay here forever.

El Rey is living witness of that. And by the way, he is still here in Comodoro, putting off the evil day when he has to face his spouse. Today he is upset because yesterday he varied his diet of whiskey by drinking a bottle of Bols gin and it made him sick, to his intense disgust.

"I can't understand it," he said just now. "I used to be a drinking man." At that moment the waiter came up to take away his two empty bottles of rye whiskey and to open a fresh bottle for this evening.

Another American visiting here now has lived in northern Patagonia for twenty-one years. He was on the plesiosaur hunt, an event famous all over Patagonia and considered as the most successful practical joke in history. According to my present informant—and his story checks so well with the facts as I know them from the other side that I am inclined to believe him—the thing was a put-up job from the start. Another American, living by one of the lakes of the Patagonian cordillera, had somewhere picked up the information that plesiosaurs were enormous aquatic reptiles that were once abundant all over the world but that became extinct some tens of millions of years ago. Being rather bored with camp life, he began to circulate the story that a strange creature had been seen in the lake, and he went on to describe it in some detail, actually describing a plesiosaur but carefully avoiding any indication that he had ever heard of these extinct creatures.

He does not deserve full credit, for the tale passed from

mouth to mouth, and as it went it gathered corroborating and convincing details. It became so circumstantial that it was taken quite seriously by a reputable scientist, and this unfortunate and humorless person actually got together an expedition and came down to Patagonia to capture the living plesiosaur. He even provided against the possibility of the creature's being killed in the struggle and brought along barrels of embalming fluid and a large hypodermic syringe. To his surprise, he found that everyone in Patagonia had a friend who had seen the plesiosaur, but that no one had seen it with his own eyes. Finally he left in disillusionment, and Patagonia has been roaring with laughter ever since.

After giving me what appears to be the inside story of this tragi-comic incident, my informant went on to say, "Now, of course that was just a practical joke, but, you know, there really is something down here. I know an Indian who lives up in the lake region, a very reliable man, too, and he says his father saw this creature. And I heard from another very good source that a whole village saw it one morning and that one man tried to catch it but was drowned. This thing is as fast as lightning and it only comes to the surface of the water at dawn. It is about forty feet long, has a long neck and a small head, a body like a turtle but without a shell, four flippers, and a stubby tail. This thing must exist, and if I were you I would capture it. You'd be the most famous scientist in the world!"

The description is that of a plesiosaur. My compatriot is very naïve indeed to think that the gag will work twice.

CHAPTER XIV

FARTHEST SOUTH

April 16.

We left Comodoro yesterday and struck out almost due south along the shore, every league that much nearer to the South Pole than we have ever been before. The road, the continuation southward of the picada, crosses several deep cañadones and climbs up over intervening high spurs from the Pampa Castillo, finally reaching a more level and practicable course on a series of flat coastal terraces. Here we had heard that there was a seal rookery, and we turned off the track to see whether we had reached it. We had not, but we had unexpectedly reached our camping place for the night. We went a few feet too near the beach, and when we tried to start up, the car simply dug itself a deep hole in the loose gravel. We worked on it until dark, breaking a jack and one block and tackle, with no success, so finally set up our beds and turned in for the night, which was clear but cold.

This morning after breakfast I walked down to some buildings visible about a league away. These turned out to be a whaling station, and I soon was able to return to the car in company with the production manager and the chief mechanic and with a bigger and better jack and other equipment. By noon the truck was extricated, and we went on to the station where we had lunch with the chiefs and looked the place over.

This is a very large and modern plant, where they use everything from the whales except the smell. The machinery is all run by electricity, generated by three Diesel motors, and the cooking steam is generated in fuel-oil boilers. The whales are pulled out of the sea up a long inclined concrete plane to the top floor of the factory, where they are cut up. The various parts are then dropped through holes in the floor down into the plant below. The blubber, meat, and entrails drop into great vats where the oil is tried out by live steam. When thoroughly cooked, the product is dried by a vacuum process, and the oil is drawn off and run through large rotary separators and then barreled. The residue from the cooking, the cracklings, goes through presses which remove any remaining oil, and the cake is then ground and sacked for feed for domestic animals and birds.

The bones are also cooked for oil, in a different part of the plant, but they are not vacuum-dried or pressed. From the cookers they go through a drying cylinder, from which they emerge as a dry powder which is sacked and sold for fertilizer. The oil from this particular plant is shipped to Liverpool where it is sold for soap and for oleomargarine. The season is just over now and the factory is not working. They are engaged in loading the products onto the mother ship which will take them to Buenos Aires where they are transshipped for Europe.

The actual whaling is done from small steam vessels under the aegis of the mother ship. The harpoons are murderous weapons with pointed, explosive heads. A swivel attaches the head to the shaft, but this joint is kept rigid until the whale is hit by trigger barbs. The long shaft or handle is slotted, with the line attached movably in this slot. The handle is inserted in the gun, with the line at the muzzle end of the longitudinal slot. When the gun is fired, the line slides back to the end of the handle and trails behind the harpoon. The grenade head is timed to explode one minute after the gun is fired. If the whale is hit, the explosion kills or disables it at once in most cases, the barbs pull out and fix the shaft and line to the whale, the swivel joint making the attachment movable and making it almost impossible for the barbs to pull out. The protagonists of Moby Dick would probably be shocked at this very unsporting way of whaling, but whaling is now a big business and not a sport. The idea is to kill the whale and not play with it. [That from the point of view of the whalers. From any broader point of view, modern whaling is brutal, destructive, and nasty. Soap can be made from vegetable oils, and butter is preferable to margarine. Unrestricted whaling will simply bring to extinction needlessly what are probably the grandest and yet least offensive beasts alive today.]

This station started as a much smaller outfit making oil from the seals still fairly abundant on this coast, and only quite recently the sealing has been given up and the factory enlarged to deal with whales.

Just after leaving the whale factory, we came to the lobería or seal rookery—in the dictionary the Spanish for seal is "lobo del mar," "wolf of the sea," but down here they are simply "lobos," "wolves," not a source of confusion, since there are no wolves (or land wolves) in Patagonia. The lobería is simply a long stretch of shingle beach, like any other Patagonian beach to the human eye but apparently very distinctive and desirable to the seals, as they congregate here in great numbers and are seen only singly or occasionally elsewhere. When we came, they were in three groups, each of several hundred animals, lumbering up onto the beach, sleeping, fighting, or just resting there in compact masses, occasionally sliding into the sea and swimming about outside the breakers. The noise was almost deafening as they barked and yelped, from the rasping basso profundo of the great old bulls to the shrill treble of the puppies. As we advanced, the herd melted and flowed down into the ocean, where they all stayed close inshore riding the surf. They would rise vertically out of the water to look at us and bark and roar, remarkably resembling a crowd of disapproving aldermen surprised in the bath.

There was one penguin in the crowd, one of the little, rather neutral penguins about a foot high which seem to be the only representatives of their highly amusing family on this coast. Even they are quite rare here, in our experience. When we first came up, he launched himself into the sea and swam gayly away, flapping his wings in the water as a flying bird does in the air. Unlike ducks and similar aquatic birds, the penguin does not paddle along with his feet, but swims in the water like a fish, using his wings (wholly useless for flight) as paddles. Later he came out again and we cut off his retreat to the sea. He then stood on his dignity and refused to run, or rather waddle, away from us but stood his ground snapping and growling at us when we came too close. Even when we touched him he refused to retreat and seemed to be intensely annoyed but not in the least afraid. When we did open a free route to the water, his fighting spirit was up and he refused to leave until we finally literally threw him into the waves.

At Caleta Olivia, farther south, there is a small, sad settlement, with a few bales of wool piled up on the open beach waiting to be lightered out to some passing ship. We had thought of staying there, as we knew we could reach no other shelter before dark, but it was all so depressing that we went on anyway. After dark we came out on a bare, flat, low pampa, almost devoid of any vegetation, however thorny, and we sped along in the night without any clear notion of where we were. To our surprise we suddenly came to a railroad, obviously the one from Puerto Deseado to Colonia Las Heras, as that is the only one within some hundreds of miles. This line was built under the same circumstances and with the same rather uneconomic idea as that from Comodoro to Sarmiento, and like the latter it merely runs inland from the coast and stops about halfway to the cordillera.

Nearby were some houses and a boliche where we stopped and learned that we were in Fitzroy. The town, if such a small cluster of buildings can be called a town, is named after the commander of the *Beagle*, with whom Darwin made his famous trip around the world a hundred years ago. Incidentally, Fitzroy's charts of this coast are still in use.

The proprietress of this establishment is not quite right in the head. One of the boys says that she has "strings in her head" and a woman here puts across the same idea by saying that "her head is boiling." We played bidou for a vermouth before supper, and she kept peeking under Coley's dice cup and telling me that he was cheating me terribly—of course she had never seen the game before and had not the least notion of what it was all about. She finally became so indignant at Coley's perfidy that she had to be led away.

April 18.

Leaving Fitzroy yesterday morning, we went down a long cañadón to the sea, and then worked back up and into another cañadón where we spent the night at the place of one Ramos, nothing to do, of course, with the Ramos of Colhué-Huapí. The house of mud, stone, and corrugated iron is very criollo but relatively comfortable and we slept well in the store-room. We had a little trouble because Ramos insisted on sealing our room hermetically, but this was solved by jamming our cots in such a way that the door could not be closed and then sitting around and wringing our hands until he gave up.

This morning we went down this cañadón to the sea and spent the day studying the geology. I think that this must be the Cañadón Tournouër of the maps, although the local lads call it Cañadón Lobo and never heard the name Tournouër. It would not be the first time that someone in Buenos Aires named a place down here and neglected to tell the local inhabitants. When we left, late in the afternoon, Ramos gave us a butchered sheep and refused to accept payment for anything, although I finally persuaded him to take two tins of fruit, which he had not tasted for many years, and to let me give a couple of pesos to his small boy. He is true native, and they are almost always very hospitable. Our few less pleasant experiences in this respect have mostly been with the Boers, Welsh, and other foreigners of Chubut, people who, with some pleasant exceptions, are not noteworthy for either hospitality or generosity. Down here in Santa Cruz the people are criollos, whom I like and with whom I almost always get along beautifully, or else English, who usually accept us as of their race and are cordial on that account.

From Ramos' place we hit the road again, dropping down

to the sea once more at Puerto Mazaredo, where a raised beach has cut off an inlet and formed a salt lake. Loomis, the American bone-digger who was here twenty years ago, called Mazaredo "the most forlorn town on the coast." Things have changed since then and the epithet no longer applies—for the reason that Mazaredo is no longer a town. Its former site is occupied by one tiny shack, not large enough for habitation, and the present "town" is a small telegraph station, nothing else, some distance inland. The title of "most forlorn" I now bestow on Puerto Visser, a thriving place in Loomis' day. Conditions have certainly not improved very noticeably on this coast in twenty years.

After many ups and downs, we reached the high pampa after dark, and followed a winding road to an estancia with the pretty name of "La Madrugada"—"Dawn." Here we were cordially greeted by an English administrator named Grant. Originally of Gloucestershire, he lived for a while in Australia, and then came to Patagonia in 1906. He has eight leagues of camp, about forty-six thousand acres, all fenced, and a very comfortable camp house, smaller than that at Las Violetas but less decayed. He is very much interested in animals, fossils, and Indian relics and has a large collection of arrowheads. He knew our colleagues Loomis and Riggs, both of whom had stopped here at La Madrugada.

We talked until late, among other things of wool hauling in the old days when the cordilleran estancias would have the big carts, chatas, make three one-way trips a year. One summer they would go down to port, back to the estancia, and then to the coast again where they would winter. The next summer they would go up, down, and up again, wintering at the estancia.

He also told us of the Patagonian cycle, the history of the average man who came out here twenty-five years ago to make his fortune. For several years the newcomer would work as a peon on some estancia, taking small wages and a share in the increase of the flocks. In a few years his share would be a thousand sheep or so, and with these he would squat on an

unoccupied piece of neighboring land. He would finance himself by negotiating a loan of two or three pesos on each sheep,
payable when the wool crop was in. On this he could live, dip
the sheep, clip them, and get the wool to market, when the
loan could be paid. By the time of the war and of the high
wool market he was all clear and rolling in what was to him
great wealth. He spent money like water, and got the habit
of spending most of the year in Buenos Aires and letting the
estancia run itself. The price of wool dropped again. He
had no savings and had to borrow on the sheep. After a few
bad years he could not pay off the loans and lost the sheep
and his land. So there he was, and now is, exactly where he
started, twenty-five years older, without a cent to his name,
and with neither the time nor the opportunity to rise again.

There are, of course, some exceptions. A few people saved
their money and left while they could. A few, like Grant
himself, stayed on but still manage to make a decent, if not
very pleasant, living here.

Still paraphrasing Grant, colonization in Patagonia took
place from the ends. From the north, settlers chiefly of Latin
origin worked down into Río Negro and northern Chubut,
while from the south came mostly British wandering into
southern and central Santa Cruz from the Falklands and
Tierra del Fuego. For a long time central Patagonia, southern
Chubut and northern Santa Cruz, where all our time has been
spent, was a sort of no man's land where outcasts and hard
characters thrived between the two waves of advancing colonization. Finally this region was also occupied, the Boer
Colony came in, oil was found, and the railroads were built.

Puerto Deseado
April 20.

We stayed two nights at La Madrugada, luxuriating in
real beds with springs and sheets, and the day between was
spent clambering up and down the sea-cliff at Punta Nava.
This morning we left and drove to Deseado, Port Desire of
the early English explorers, along a very uninteresting road,

almost perfectly flat pampa. Just at Deseado there are some
hills of hard porphyry, the same rock that outcrops up at
Bustamante, on the north rim of the Golfo de San Jorge.
Now we are south of that enormous gulf, having worked
along its entire shoreline.

Deseado owes its location to the Río Deseado, which here
empties into the sea. The river is not navigable, is, indeed,
a mere trickle of water, but its mouth is a wide estuary into
which ships can come and where they find some shelter, rare
enough on this coast. The settlement is very old, was one
of the first in Patagonia, and continues to be of considerable
importance. I do not know, but from its appearance I would
guess that the town had a population of around two or three
thousand, which makes it a large city here. In recent years it
has been outstripped by the relatively very young town of
Comodoro, but this is largely because of the oil fields near
the latter. If or when the oil gives out, Comodoro will doubtless fall back to the importance of Deseado, or perhaps less.

The town is the usual collection of tin houses in a bleak
landscape, a Comodoro on a slightly less ambitious and progressive scale, and seems to us to be very uninteresting. The
social life among the non-Argentine here appears to an even
greater extent that that of Comodoro to consist of a continuous
series of Gargantuan drinking parties. I trust that we can
leave quickly and quietly. At present the car is being repaired,
a constant necessity.

Comodoro
April 24.

We left Deseado safely after only one night there and
drove out along the railway as far as Fitzroy, twenty-eight
leagues, where we spent the night at the establishment of
the lady with strings in her head. The trip was marred by a
tragedy. Having broken our jack trying to extricate the car
from the gravel near the whaling station, we purchased in
Deseado a new, very large, and handsome jack of the kind
used on the big wool carts. Not having time to fasten it else-

where, we loaded it in the back of the car. The road is very rough, as are all roads in Patagonia, and when we arrived here this evening we found that the jack had bounced onto our "pichería," the box in which we keep our pet armadillos, and had killed our favorite Florrie.

The next day we studied some rocks near Pico Truncado, farther along on the road from Deseado to Las Heras, spending the night at a boliche in the town of Pico Truncado. In the morning we cut straight across to Caleta Olivia, finding the country very boring. The whole region there north of the Deseado River is barren, flat pampa. At Caleta Olivia we had lunch at the boliche of Max Kleine, a German sailor who has oddly come into harbor as an inn-keeper there. He possesses a few words of English, which he employed, while we ate, to praise the German commerce raiders and their war activities. Hence without incident we retraced our steps over the very bad main road to Comodoro, which we reached late last night.

Now we have about a week's work getting our fossils packed and shipped off to Buenos Aires, copying notes, writing reports, making up accounts, and doing the many other things incident to winding up an unusually long field season. Most of our fossils and equipment will go to Buenos Aires on an oil tanker. We will take what we need on the road and a few of the fossils and will try to drive up.

Except for brief observations on the way north, our field work is done. Winter is upon us. It has been a very successful season and, on the whole, an interesting one. But I am glad that it is over. It will be an infinite relief to get in out of the wind, and to see the lights of Buenos Aires again.

CHAPTER XV

FAREWELL TO PATAGONIA

PATAGONIA was not through with us. There were times on the long road north when we felt as in those nightmares when you try and try to run and the surroundings seem to cling viscously and to hold down your pace to a bare crawl despite desperate efforts. There was that first day out, when dark overtook us before we could reach Nollmann's boliche. The road was dry and dusty, but beneath the dust there was a hidden spring and the car was soon deeply sunk in foul alkaline ooze. We worked there heart-breakingly until late at night, and finally reached the boliche and rest, more dead than alive.

The Nollmanns were pathetically and charmingly glad to see us. When we left the next morning they loaded us down with home-made sausages and other delicacies, even including a pickled ostrich ham. We also saw some of our Río Chico acquaintances. There was an old criollo there whom we knew; six months ago he had a bad cough and the maternal Señora Nollmann put a large mustard plaster on his chest. We hoped he was better and asked how the plaster had worked. "Ah, magnífico!" he said, and bared his chest. There were the tattered and filthy remnants of the plaster, still bravely hanging on.

Back on the picada and headed north, we stopped for a moment to say goodbye at the Hotel de La Pampa, but sailed by Malaspina without a word. Henceforth the road was new to us, new but quite the same. The Meseta de Montemayor is a continuation northward of the Pampa de Castillo and is indistinguishable from it: nearly flat, occasional undrained depressions with alkaline deposits, clafate and malaspina bushes, now and then an ostrich or guanaco, and many martinetas. Just as it was getting dark, we reached Lochiel, the estancia where Townsend, a friend met in Comodoro, is administrator.

This estancia Lochiel is a grand place, incomparably the best that we saw in Patagonia. The house has rugs, books, comfortable chairs, and running hot and cold water. It is surrounded by trees, poplars and firs, and there are a tennis court, a lawn, neatly trimmed hedges, and flowers. La Madrugada, hitherto the most luxurious place which we had seen or could imagine, is positively old Patagonian in comparison with Lochiel, a great triumph of man over matter. The feeling of being in England was rather spoiled the next morning when we looked out and saw the poor trees whipped by a terrific wind, and beyond the little patch of carefully nursed verdure saw the great columns of sand rising in the gale and the same bare barrancas. Such an oasis does not demonstrate the potentialities of Patagonia so much as it does the perseverance of man, and in a way it only accentuates the general misery of the country. [That is in part a panegyric to Lochiel penned when I was there. Now I can see how distorted my perspective had become by my stay in Patagonia. Lochiel's superiority to its surroundings remains a tribute, but in fact it is less comfortable and very much less attractive than the average modern farmhouse on a progressive small, American farm.]

From Lochiel, on the morning of May third, we went back to the picada and continued on it toward Rawson, the territorial capital. In three days of travel on Patagonia's only main road, we passed just four automobiles and three wagons; in the first hundred miles after leaving Lochiel, we saw only one house. This is the ideal spot for those who love solitude and space. The country is the same as that seen before—flat pampa. We passed one small group of thorn-covered hills, and just before coming to the Chubut Valley the surface became a little more rolling, with the road going up and down a series of long gentle slopes.

At the edge of the valley the road goes on to Rawson, which can be seen in the distance, but we turned off to the left and went into the town of Trelew. Here again we thought that the savage Patagonia we knew was past, a recurrent thought and one which always proved false until the moment when we left Patagonia for the second time. In those surroundings the Chubut Valley is idyllic. It has an alluvial flood plain which can be irrigated from the river, and the colonists have turned this narrow strip of soil into a garden spot. Imagine seeing a placid river, willows drooping on its banks, a quiet country lane, and a rustic bridge. Our hearts leapt at the sight, and we quite overlooked the fact that the water was muddy, the trees spindling and wind-torn, the lane almost impassable, and the bridge in ruins.

Trelew is a smaller town than Comodoro, but very much more attractive, with its plaza and its trees and grass—very inferior trees and grass, to be sure, but still green and growing and a delight to see. Our hotel was a magnificent structure with three stories, tile floors, running water, and even a little patio with a garden. We stayed there two nights, as the car was laid up again, and I improved the intervening day by driving up the valley with young Dean to examine some reported fossil deposits.

We had met Dean, senior, in Comodoro and he was very cordial. He is headmaster of an English school and also looks after two Anglican churches as layman. His school, where instruction is in English, has about forty boys, English, Welsh, and a few Argentines. Young Dean attends the National High School, where he hopes eventually to win a Spanish scholarship for Oxford. His instruction is in Spanish, or in Argentine, and he speaks it like a native, although his father speaks it badly and his mother hardly at all in spite of many years in the country.

Dean told us something of the Welsh Colony of which Trelew is the chief center. They came out sixty-five years ago, when this region was completely wild and isolated, coming with the idea of escaping English persecution, so they say, although they seem rather vague on the subject. They decided to form a little Wales, to speak Welsh only, to have nothing to do with anyone else, and to be a world unto themselves. On land provisionally allotted by the government they placed their farms, dug ditches with pick and shovel, planted trees and sowed fields, and built houses and barns. They specialized, and still do, in dairy products, particularly in Chubut cheese which they have made famous. All their affairs were managed by a Mercantile Company which they formed among themselves and which served as store, market, bank, and godfather to the whole colony.

Younger generations growing up in almost complete isolation, partly because of the ideas of their fathers and partly because of their geographic position, were naturally poorly fitted to cope with the outside world when they did come in contact with it. As the valley prospered, such contacts became inevitable. The government began to take an interest in what it had given as waste land but what was now fertile fields. Argentines and foreigners appeared in the valley and the colony lost its purely Welsh character. The Mercantile Company scorned to deal with any English banks or capitalists and took its mortgages, when necessary, to the Banco de la Nación, which (doubtless quite advisedly) gave very stringent terms. At the time of our visit the Company was heavily in debt, paying tremendous interest costs, and was on the brink of insolvency. It seemed very imminent that the Welsh would lose their colony as such and their dream altogether. [I became very tired of these stories of hard luck and of failure in Patagonia. I longed to meet just one person who liked Patagonia and had made a great success there, but I never did. The nearest approach (and not a bad one) was among the many puesteros who are poverty-stricken, certainly failures in any financial sense, but who keep alive and are content with their lot.]

Dean, junior, who, I am sorry to say, displayed a hearty scorn for his Welsh neighbors, told me, and swore it true, of one who used to hunt ducks. He hunted around until he found a lake where they congregated and where he could get two or three, sitting, with one cartridge. This was too expensive for him, so he spread grain and when they came to eat it he could kill five or six at once. This still seemed too costly, as cartridges do come high there, and so he bought some cheap alcohol and soaked the grain in it. Then he could get all the avian inebriates he wanted by simply knocking them over with a stick. After a while, however, he began to worry so over the cost of grain and of alcohol that he gave up duck hunting altogether.

Young Dean also assured me that the Welsh are very jealous of their language, feeling that it gives them their one advantage, and that they are determined that no one else shall learn it. He said that a young Welshman of the modern generation and ignorant of the language of his forefathers was sent out from Wales to take charge of a chapel in Trelew. He found that his congregation could not or would not understand him if he spoke English or Spanish, and decided to make a desperate effort to learn Welsh. No one, however, would help him to learn the language or give him practice by conversing in it with him, and he eventually had to go back home.

From Trelew we drove to Puerto Madryn, a small and typically Patagonian coast town where the Argentine naval training ship, the frigate *Sarmiento*, was anchored, and hence along the coast to Puerto Lobo, which consists of a police station and a boliche. From Madryn to Lobo the road was good, the best we had seen since leaving the United States. At Puerto Lobo, however, there had been a tremendous tide the previous day, exceptional even on this coast of amazing tides. "In a century," said a genial bolichero, "you will not see another such tide," and "No," said I, "I hardly expect to." The result was that a large section of the road had been taken out to sea and we had to make a long and hazardous detour inland. Finally we got back on the picada which was here excellent again except that it had also been washed out in spots. We would be rolling along at sixty or seventy (kilometers, I mean) and suddenly there would be a deep canyon right across the nice smooth road. At least once it seemed certain for a sickening moment that we would not live to take our collection back to New York, but we got through safely with only two punctures and one blowout for the day—by this time our tires had seen so much cruel and unusual punishment that they were rapidly going to pieces under us.

Between Puerto Madryn and San Antonio Oeste one leaves the territory of Chubut and enters that of the Río Negro, which together with Neuquén to westward forms the northern part of Patagonia. The country is still quite Patagonian, but differs enough from that farther south to look unfamiliar after a sojourn in southern Chubut. The hills are lower and more often form series of low, broad, rolling ridges than the terraces, mesetas, and high pampas of the south. The vegetation, too, is different. From Trelew to a point well over halfway to San Antonio much of the country is covered by a growth of scraggly bushes four to six feet high and the ground below them is bare and pebble-strewn, with few or no smaller bushes or clumps of grass, even of the wiry sort sometimes found on the dry pampas.

After sunset on the fifth we reached San Antonio Oeste, having made the unusually long run of four hundred fifteen kilometers from Trelew in one day. The town is small and bedraggled, half corrugated iron and half brick, occasionally stuccoed. Such importance as it has is due to its being the railhead. From here it is possible to take a train to Buenos Aires. In spite of the existence of much transient trade, we found the chief hotel below the best Patagonian standard.

Having our tires fixed delayed us until noon the next day, and even then we had much trouble finding our way out of town because the main road had been washed away by the high tide of two days before. When we did get back on the main road, we could hardly believe it, for we found only an ungraded trail through a succession of sand dunes and deep mud holes. At the fences it had gates instead of the

much more convenient cattle guards usual on the main highway. Another inconvenience was due to the very large
tracks in which freight is hauled along that route. They leave
deep ruts which are farther apart than the gauge of a normal
car so that we could not drive in them and yet were continu-
ally having two wheels fall into a rut, leaving the other two
high and dry and our top-heavy load inclined at a precarious
angle.

Over this sort of going, it took us four and a half hours
to make a hundred kilometers, about sixty miles, and by the
time we reached the first settlement it was too late to think
of going on. This little town, Conesa, is even more verdant
than Trelew and the presence of gum trees, which require
a warmer climate, reminded us that we were moving nearer
the equator. In fact, Conesa is almost exactly on the fortieth
parallel, the latitude of Philadelphia, but in general Patagonia
has a more severe climate than that of corresponding latitudes
in the northern hemisphere. The surrounding country is very
barren and is for the most part covered with thorn bushes,
up to ten feet high, in otherwise almost completely sterile
soil.

We have at various times had a dozen or more tame pichis
and our hopes of taking one back with us had been high. One
by one they fell by the wayside, but we still had one when
we reached Conesa. We left this to spend the night in the
cab of the truck, and the next morning it had been stolen,
doubtless to figure in the daily menu of some Conesan. So
after all our efforts we came back pichiless.

We got away for an early start, with the idea of reaching
Bahía Blanca, another idea that would have been excellent
if it had worked. Just beyond the town we crossed the Río
Negro on a ferry. The river valley itself is very pleasant,
fertile and grassy and with groves of large trees, but immediately beyond it the gravel and sand and thorn bushes start
again. This type of country continues without a town and
with only two or three poor houses for a hundred miles.

Then we reached the end of Patagonia. Another river

valley appeared, and beyond it an abrupt and continuous bluff. It was the Río Colorado, and the bluff the southern edge of the Territory of La Pampa. While Patagonia has no legal boundaries, this is generally accepted as its northern limit.

We stopped in the town of Río Colorado for lunch and to look over the car, which proved to have a broken spring as usual. This delayed us until about four, when we drove off, still expecting to make Bahía Blanca by eight or nine. As we crossed the Río Colorado, on an excellent bridge, we waved and happily cried "Adiós, Patagonia!" We should have said "Hasta luego!"

After climbing the bluff we came onto a level road, formerly worked but then neglected. However, it was dry and we rolled merrily along. Probably imagination accentuated the difference, but immediately on leaving Patagonia the country seemed wholly changed. Although on a pampa without running water, there were trees, small but real, and soon we began to pass fields under cultivation. Even the hitherto omnipresent Patagonian pebbles disappeared just south of the Río Colorado and did not reappear. This new country very closely resembles the high plains of eastern Colorado in the United States, a resemblance enhanced by the presence of innumerable tumbleweeds, or Russian thistles, rolling along the road and piled high along the fences.

Six leagues from Río Colorado we were sailing along at top speed when there was a loud bang announcing that a steering knuckle had broken. By all the laws of physics, the car should have turned over and smashed us flat, but our usual luck held and we did not even leave the road. We resigned ourselves to an all-night bivouac, but luck was still ours and in little more than an hour, shortly after dark, three young lads drove past in a car. I flagged them and went in to Río Colorado, leaving Coley with the truck. Everything was closed up in town, but after much searching I managed to find and awaken an automobile agent. He did not have the part we needed, but improvised one by sawing bits from another part. With this precious object in hand, I hired a car and again drove across the bridge, this time saying in even more heartfelt manner and with truth *"Adiós Patagonia!"*

It was a wild ride. The engine of the car was literally tied together with string. The motor stopped once and the driver got out, untied a string, unwrapped a long rag bandage, looked in perplexity at the wound, rebandaged it, tied it again with a string, cranked the car, and it went on without further trouble. We stopped again about three leagues from town to give a vermouth bottle full of gasoline to some stranded wayfarers with two cars. Although their tanks were dry, they hoped to get the lighter car into town on this bottle. Then on we roared through the dark, leaping madly from side to side of the road, sometimes jumping right out of the road and bouncing back onto it. So we brought the repairs from Río Colorado to wherever it was that we were stuck.

I thus had two trips to Patagonia, one of over seven months and one of an hour.

Coley was sound asleep, not expecting me till morning, but roused and soon had the makeshift part adjusted. We went back a few miles to Anzoategui to spend the night.

That Anzoategui is a queer place. It comprises chiefly one row of stucco buildings, the central ones very florid and resembling an amusement park, all in an advanced stage of disrepair. Together with many leagues of land, it belongs to a large and once prosperous corporation. A huge sign, barely legible because so many letters had peeled off, proclaimed that here anyone willing to work will find a job and happiness, that there are no policemen because all workers are united in a great brotherhood, that the Ferrocarril Sud (which railway runs past the place) helps those that settle along its route, and that so affirms Fortunato Anzoategui, president of the company. The colony is supported (so said the innkeeper, but the support did not seem to me to be brilliant) by salt manufacture, woodcutting, and agriculture.

The wood is now almost all cut, and they are turning to wheat growing. The property is really enormous and even has its own narrow-gauge railway, with many kilometers of track. There is a very low-class inn, a butcher shop, a large machine shop, two or three stores, and so on.

The machine shop enabled Coley to make our repair more permanent, and we were able to get away early on the morning of the eighth. For a change, all went well and we reached Bahía Blanca without mishap.

Bahía Blanca definitively marked our return to civilization. We had been approaching it little by little, and every time we had seen a tree or a decent house we had exclaimed "Civilization!" Yet the real thing burst on us unprepared. Four-story buildings! Paved streets! A plaza with palm trees! Street cars! A hotel with an elevator and running hot water in the rooms! We were so overcome that we went into a barber shop and were shaved. It was a wrench to part with my beard, and my face emerged shyly from its seven months in hiding.

We never did learn not to set ourselves definite goals. We left Bahía Blanca after a luxurious night, determined to reach Maipú, and failed miserably. The road soon became a slough through which we wallowed. The Argentine government was formulating a very ambitious plan of road work when I left there, and I am quite willing to believe that by the time these words are published things are quite different. But when we were there the roads were a black disgrace to a civilized country, undoubtedly the worst to be found in any nation otherwise so progressive. In Patagonia we were in a frankly wild country with only a few small pioneering communities and good roads were not to be expected, but here we were in a prosperous and civilized region, passing through one of the richest wheat areas on earth, on a main road between some of the principal cities of the Republic, and that road was simply a lane filled with almost bottomless liquid mud. Even in the immediate vicinity of Buenos Aires, I did not see a single road in the Argentine which could be called fairly good from a motorist's viewpoint. They were working on a concrete highway from Buenos Aires to Luján, an hour's drive or less, and that was the only hopeful prospect, although in the last year conditions have probably improved there.

We got through, but at the expense of such effort and so slowly that we fell far short of Maipú and had to stop at Juárez; the trip to Maipú which looked on the map like a short and easy day's drive took two long days of the greatest difficulty and physical effort.

The countryside, however, was prosperous, green, and beautiful. The land is gently rolling, with very slight relief but nowhere really flat. Most of it is cut up into broad wheat fields, in some of which the grain was sprouting, while others were being planted and still others were being worked by tractors preparatory to sowing. The unplowed land has excellent grass and there are many cattle and some sheep.

The road beyond Juárez next day was even worse than what had gone before. In one place we had to pass a bridgeless arroyo (as they call a wet zanjón in those parts) down a steep bank, through water and mud deep enough to enter the engine, and up another very steep and slippery bank. By pushing and luck, we had just made this when we bogged down completely and almost hopelessly in bottomless black mud. Fortunately another truck came along after a while and pulled us out, so that we were able to lunch in Tandil.

The country around Azucena and Tandil is very attractive. There is a zone of fairly high, rocky hills, the last to be seen on the road to Buenos Aires. The broad intervening valleys are green and fertile and are enlivened by rows of poplars, willows, and other trees. It is a grazing country, producing meat and dairy products, in contrast to the wheat belt traversed by us the day before, but there are also farms where corn and other agricultural products are raised. The hills give rise to quarrying, chiefly for paving stones, as hard rock is a thing unknown north of there on the great pampa.

That evening an hour after dark we did reach Maipú, which had begun to seem like a will-o'-the-wisp, and with no further mishaps than our daily puncture and the failure of our lights so that we drove into town by the touch system. The hotel there is fairly typical of a provincial town. The square patio is surrounded on three sides (in other cases often on all four) by a covered walk. On one side are the common rooms, and on the other sides the bedrooms, each opening separately onto the patio. The building is one story high, of stucco, with tile floors on the walks and the common rooms and wooden floors in the bedrooms. One or at most two bathrooms serve the entire place, and they are reached by crossing the patio. The bedrooms have very high ceilings but small floor space. Those fronting on the street have high shuttered windows, and the others have no windows, ventilation being considered positively undesirable. A great veil of mosquito netting is fastened to the ceiling and completely envelops the bed.

[That is written in large part from memory. In my journal I find the entry, "The hotel is probably all very picturesque, but if so I have been in the country too long to see it with foreign eyes. It was quite ordinary and rather low-class accommodation for the night." One thing that impressed me very much when we came back into the provinces from Patagonia was that everything seemed very commonplace, aside from the thrill of seeing green and fertile land again. I had to keep reminding myself that I was still in a foreign country. Unlike that of Patagonia, usually very distinctive, the provincial landscape could be anywhere and I found that it was only when I made a particular point of it that I noticed the peculiarities of buildings, dress, customs, and speech.]

From Maipú onward, the story of travel is short. There we intersected the road from Buenos Aires to the resort town of Mar del Plata. We had been looking forward to this, as it was famous far and wide as the Argentine's good road. It proved to be better than the average, but still bad. Only a few short sections had been graveled and the rest was a graded dirt road, rough when dry and hopelessly muddy when wet.

The country is very dull, simply the endless and perfectly flat pampa. Along the road there are straw huts, built of withes, plastered with mud, and straw-thatched, picturesque and dirty, as most picturesque places are. Here and there are large estancias with comfortable big houses, sometimes with large lawns, groves of trees, and even formal gardens. Towns became more frequent as we advanced toward the capital. Finally on the afternoon of our eleventh day from Comodoro we reached La Plata, where we were cordially greeted by Doctor Cabrera and where we spent the night.

On May twelfth we drove into Buenos Aires. Going through the double police cordon which, for some reason, is maintained around La Plata we picked up a young policeman with a bad cold and gave him a lift in to Avellaneda, a suburb of Buenos Aires. I had hoped not to be in such close contact with the police again, and had not expected it to be so little unpleasant. In the outskirts of the city we went around a few unnecessary circles and tried to cross a one-way bridge the wrong way, but soon found ourselves in the center. We had promised ourselves the triumph of riding down the Avenida de Mayo and ending our long excursion in front of the Casa Rosada, but we made the scientific discovery that trucks are not allowed on the Avenida. Undaunted, we slipped down a side street and came out in front of the government house nevertheless, driving grandly up to the front door in the teeth of two magnificent sentries. Being for purely decorative purposes, they ignored us, but several common or garden sentries began converging on us with fixed bayonets, not sharing the expressed opinion of the onlookers that we must belong to the army. Our goal was achieved, and we departed.

There are two more long stories of the Scarritt Expedition, but I shall not write them now. One would tell of the several months that I spent in Buenos Aires and in La Plata studying their fine collections of fossils in the museums there and enjoying Argentine life at its best. The other story would tell of the weeks spent in obtaining permission to export the fossils, of the difficulty of getting everything in order in time, of our final success, and would perhaps go on to tell of the still longer, still far from complete work, preparing and studying the collections in New York.

Our further troubles with the collection in Buenos Aires were not due to lack of coöperation from the authorities, still less to any actual opposition from them. The struggle was with the red tape of bureaucracy, with the forms and formulas that had to be followed, and with a long list of official requirements all keyed to a slow tempo which we were trying desperately to accelerate. The officials themselves were uniformly courteous and some went to great lengths to help us in every way possible. Almost without exception, that has been my experience with the citizens of Argentina, whom I sincerely like and admire.

June sixth dawned. The last requirements were met, the last of a mountain of papers stamped and signed. We rushed the fossils to the dock without a moment to spare. As our last box was swung up onto the deck of the S.S. *American Legion*, the ship was already under way. I stood on the dock and waved goodbye to Coley and to the fruits of our long effort.

Se acabó!

AFTERWORD

This book was written about a half century ago, in 1932 and early 1933, and was first published in 1934. It is a true adventure story, some parts abbreviated and some elaborated from my journal of an expedition to Patagonia in 1930–31. Mainly in chapter 4, its tells some of the reasons for our being there and how we did our work. The book as a whole is about what we did in Patagonia during our months there on that one expedition. It also says a good deal about how I felt about the things we saw and experienced.

The account starts abruptly with a revolution in Buenos Aires, and chapter 1 has been likened to "a Graham Greene thriller." It annoyed some of my Argentinian friends who felt that I had strenghtened the stereotype of banana republics and Latin-American misrule and disruption. All I could say then, and now repeat, is that this revolution—a rather mild one as such things go—did occur exactly when and how it is here recounted. It was my first introduction to a Latin-American country and could not but be introduced as the beginning of my adventures there, adventures that became quite different after the brief revolution and during our subsequent travels through Patagonia.

The original text, reprinted in full in the present volume, went right on from that exciting beginning without a backward glance at the origins of the expedition, without clearer identification of some of the

296

people involved, and without a forward glance at some of the things relevant to the book that happened later. I do feel that the book as originally published was and remains a unified whole. Nevertheless, a number of those who have read it have found their curiosity aroused in several respects, and the purpose of this Afterword is to try to satisfy such curiosity. With a different perspective, my younger colleague Larry Marshall has added to the present edition some discussion and evaluation of the work in further and wider approaches to the subjects here introduced.

One of the things I did not think to say in the original edition is that having been born on 16 June 1902, I was twenty-eight years old throughout the whole of this particular Patagonian expedition. As I am now eighty, that naturally appears to me an almost ridiculously early age, especially for setting out on so difficult a task in so wild a region. At the time, of course, I felt quite able to accomplish this, and in retrospect it seems that the outcome did justify my youthful self-confidence.

Before this expedition I had both trained myself well and been trained well by others. Three years as an undergraduate at the University of Colorado, then one year as an undergraduate and three as a graduate student at Yale had given me a sound basis in both geology and biology and had focused my main interests on paleontology, especially on vertebrate paleontology, even more specifically on the paleontology and history of mammals, the class of vertebrates in which we humans have a numerically small but uniquely peculiar part. Thinking historically, it seemed most logical for my study of mammals to start at the beginning. It happened that in the United States the Peabody Museum at Yale had the major

collection of the oldest known mammals, those con-
temporaneous with dinosaurs in the Mesozoic era.
With the somewhat hesitant consent of my major professor, Richard Swann Lull, who was also director of
the Yale Peabody Museum, I wrote my doctoral dis-
sertation on the American Mesozoic Mammalia, and
with some revision it was published as a large memoir
by Yale University Press. A postdoctoral year, supported mainly by the National Research Council and
spent mostly in the British Museum (Natural History),
but with visits elsewhere in England, France, Swit-
zerland, and Germany, produced a memoir on all the
other then known Mesozoic mammals. This was published by the British Museum (N.H.).

While still a graduate student I had acquired experience in fieldwork, collecting fossil mammals and
making proper field observations as to the geographic
and geological data necessary for eventual scientific
study of the specimens found and collected. This ini-
tiation was for me the best possible, as it was spent
primarily as field assistant to William Diller Matthew
of the American Museum of Natural History in New
York City. W. D. Matthew in 1924 was the leading
student of fossil mammals in the world, 33 years my
senior in age and light years away in experience and
knowledge. In 1927 he left the American Museum to
go as professor to the University of California at
Berkeley, where he died in 1930 at the sadly early age
of fifty-nine. In the meantime, on Matthew's recom-
mendation in 1927 I was taken in at the bottom of the
curatorial staff of the Department of Vertebrate
Paleontology in the American Museum when Matthew left from the top. In 1928 and 1929 I started some
fieldwork on my own for that museum, first in Florida
and then in the San Juan Basin of New Mexico.

The reader may by now be wondering what connection all this has to do with the story of "attending marvels" in Patagonia. I will soon make that connection clear.

With my two monographs on Mesozoic mammals I had worked myself out of a subject for fieldwork, research, and museum exhibition; I had studied, described, classified, and discussed all the earliest mammals then known. Many more Mesozoic mammals have been discovered since then, but for a few years I had to look elsewhere. As I had studied all the then known Mesozoic mammals, those dating from about 190 million to about 65 million years ago, a span of about 125 million years, it was logical to turn next to those from the first two epochs of the age of mammals, the Paleocene and Eocene, from about 65 to about 40 million years ago.

Although Matthew had also worked on a great many other things, much of his attention throughout his professional life was focused on the mammals of those two epochs. For some weeks on the field trip of 1924, Matthew and I had shared a room in a boarding house in a village in the panhandle of Texas. In the evenings we often talked about the faunas of those ages. Matthew remarked that there was, as always, more to be learned everywhere, but there was fair progress being made on this subject in North America, Europe, and Asia. (This was due mainly to him, as regarded North America especially, which his work made evident but which he modestly did not stress.) He told me that someone in Argentina named Florentino Ameghino, who had died in 1911, had published voluminously on some ancient mammalian faunas from the region of Argentina called Patagonia which did not seem to fit in with those known from other continents or with likely determination of their ages and relationship. Problems existed there that were well worth tackling both in the field and in the laboratory.

That stuck in my mind and was reinforced as I became acquainted with William Berryman Scott. Scott was a mammalian paleontologist who was professor of geology at Princeton University for many years. He was well into his sixties when I met him and was sixty-nine when I joined the staff at the American Museum, where he often visited. From 1896 to 1899 John Bell Hatcher, a skilled collector of fossil vertebrates, had worked for Princeton making a large collection of fossil mammals from Patagonia, and Scott took over the task of editing and in good part writing voluminous quarto reports on that collection. He had also visited Florentino Ameghino in Buenos Aires and studied parts of the Ameghino collection, but he did not visit Patagonia or make any Argentinian collections. He liked Ameghino personally, but they disagreed strongly on many points. The whole Hatcher collection was of much later age (Miocene in geological terms) than that then of most interest for me, but Scott also wrote a general history of South American mammals in his classic *Land Mammals of the Western Hemisphere*. In our frequent conversations, Scott urged that the older faunas and fossil beds should be reinvestigated, especially as the collections described and interpreted by Don Florentino had not been collected by him but by his younger and less literary (but no less intelligent) brother, Don Carlos. Scott strengthened and strongly recommended my desire to follow in the footsteps of the Ameghino brothers. He lived to know that I did so. He died in 1947 at the age of eighty-nine, after my expeditions and a number of my shorter publications on the early mammalian faunas of Patagonia, but a year before the first of my major monographs on them.

Another factor leading to the first Patagonian expedition was the interest shown by Friedrich Freiherr von Huene, then professor of paleontology at Tübingen University in Germany. He was primarily interested in dinosaurs and other reptiles and had visited museums and collecting localities in Brazil and Argentina. In the latter country, he was impressed by the early fossil mammals from central Patagonia, and he proposed that there should be a joint expedition from Tübingen University, represented by one of his students, and the American Museum of Natural History, represented by me. This crystallised my growing desire to organize a Patagonian expedition, and I drew up a plan and submitted it to Henry Fairfield Osborn, who was then president of the American Museum and also de facto head of its Department of Vertebrate Paleontology. He heartily approved the plan, but carefully specified that the museum could not support it financially (although he and the other trustees were all wealthy men as wealth went in those days). At about the same time, von Huene wrote that he and Tübingen, too, could find no money to pay for their proposed part in a Patagonian expedition. That left it up to me to find some money or to forget this project.

At this point aid came from an unexpected and surprising source. A young man of about my own age was interested in the museum and through his family was acquainted with some men in financial circles. This was Coleman S. Williams, the "Coley" of the narrative. Through his family connection he knew and introduced me to Horace Scarritt, a Yale graduate (like me, but at an earlier time and on a different social

level) who had been a vice-president of Bonbright and
Company and had been canny enough to obtain and
retain what was then a considerable fortune. He
finally agreed to finance a Scarritt Patagonian expedition under my command and with Williams—
Coley—as the only other North American member.
This took some persuasion and more than a few
drinking bouts. My perhaps tactless parody of Nathan
Hale's supposed last remark was "I only regret that I
have but one liver to lose for my museum."

It is to be remembered that our project was planned
on the brink of a world depression, and also that there
was then no National Science Foundation or any other
government source for funds for museum expeditions.
There were also no funds for such out-of-house expenses in the museum's own budget.

What with talking a company out of a car rather like
a pickup and purchasing other materials as cheaply as
possible, we got together a camp and off-road travel
outfit. So, as related somewhat parenthetically in
chapter 1, we set off from New York on 8 August 1930
on the ship *Western World.* Mrs. Coleman Williams,
Dora, accompanied us on the ship to Buenos Aires and
returned home alone when Coley left Buenos Aires to
join me in Patagonia at Comodoro Rivadavia. Dora
and Coley were married shortly before we left the
United States, and the voyage to Buenos Aires was a
sort of honeymoon for them.

On 26 August, we landed in Buenos Aires and went
to an old, inexpensive, but pleasant and comfortable
hotel where we engaged a suite with a double room,
single room, reception room, and bath for the peso
equivalent of $14.80 per day for three people and
three meals a day. This was the old, now extinct Hotel
Jouston, on the central point of the Plaza de Mayo. I

record it now because it gives a hint of things long
past, and because in the interests of brevity and in
order to have a more dramatic lead-in, chapter 1 starts
with 4 September, the day when the revolution actually began, although in Argentine history it is called
the "Seis de Setiembre" (6 September), when the
shooting began.

There are many people mentioned in my journal
whom I did not think it necessary to name in the book.
Those who are named in this book for one reason or
another do not seem to need further characterization
or explanation here. Clearly, apart from Coley and me,
the principal figure in the story is Justino Hernández,
and his part in our venture is fairly clear. However, I
should have mentioned Casimiro Slapeliz, a Lithuanian of long residence in Patagonia, bright and
knowledgeable, including knowledge of what fossil
collecting is about. In the most friendly way he introduced me to a nephew of his, ordered the nephew
to go along with me and take care of me, then ordered
me to take the nephew along. We both obeyed. The
nephew was Justino Hernández, and Don Casimiro is
the brother of Justino's mother. I say "is" because
now, more than fifty years later, Don Casimiro is still
living and by all accounts quite sprightly, although he
acquired planes, one after the other, that he flew himself, wrecked, and walked away from.

The book ends with Coley sailing away for New
York with our collected fossils, now packed and
crated, while I stayed in Buenos Aires. In this Afterword I add brief notes on later events relating to the
first Scarritt expedition.

First, I stayed several more months in Argentina
over the southern winter, which is decidedly cool in
Buenos Aires, well south of the tropics, but far from

being as cold as winter in New York. My time was
divided between the two great natural history
museums of Argentina, much the most important in
South America and conveniently, although rather
oddly, only an hour or so apart by train or by car. The
Museo Nacional de Ciencias Naturales "Bernardino
Rivadavia" was and still is in Buenos Aires, although
not now in its original location. The Museo de La
Plata was and still is associated with the Facultad de
Ciencias Naturales of the Universidad Nacional de La
Plata, and as in 1931 is still in its own impressive
building in the park which is essentially the campus
of that prestigious university in the city of La Plata.
(During Perón's rule over Argentina, the city, the uni-
versity, and the museum were all renamed "Eva
Perón," but they have happily reverted to their original names.)

The great fossil collection of the Ameghino brothers
had been acquired by the nation and housed in the
Museo Nacional, then in tight quarters in an old
building in the Calle Perú just off the Plaza de Mayo,
where the Casa Rosada ("Pink House"), essentially
the capitol of Argentina, is situated. All of the types
and many other specimens of the many genera and
species of fossil mammals collected by Don Carlos
and described and named by Don Florentino were
housed in that museum, and I spent many weeks there
studying and making notes on the Ameghino specimens from the two major faunas of which I had now
also made collections. I intended to work over these,
revise their classification and nomenclature as far as
necessary, and then write a monograph based on all
the known fossils of those faunas. Lucas Kraglievich, a
noted paleomammalogist, was then in charge of the
Ameghino collection. We became good friends and he

facilitated my work and also made my days at the
museo pleasant. The director of the museo, Martín
Doello-Jurado, was also helpful and friendly. He arranged to have all the important Ameghino specimens
that I was studying photographed and had copies
given to me. I was staying part of the time in a boarding house and part in a hotel, both near the museo,
and outside of working hours I had some pleasant social contacts with both resident Americans and
Argentinian friends.

The highlights of my stay there both in a social and
in a scientific sense were the many afternoons when
Carlos Ameghino came to the museo and we talked
and drank yerba maté together in the gaucho fashion
most acceptable to both Don Carlos and me. (This
beverage and the gaucho way of serving and drinking
it are described in chapter 4; it is also much drunk in
Buenos Aires but usually in a nongaucho way, even to
sipping it from tea cups.) Don Carlos had a severe
chronic ailment, which was later fatal, but it was my
good fortune to be there when he had a remission and
was often able to spend an hour or so sitting with me
at the museo. He had spent sixteen field seasons in
Patagonia in the 1880s, 1890s, and early 1900s, when
the region was even wilder and travel very much more
difficult than in the 1930s. He kept his field notes in
his head and only rarely on paper. To an astonishing
extent, they were still in his head in 1931 when he
was a sick old man. He remembered where in a then
virtually unmapped region almost each one of his
specimens had been found.

In La Plata, which is the capital of Buenos Aires
Province, there were and even now are no good
hotels. However, I was installed in an internal guest
room in the museo quite comfortably, although meals

were sometimes a problem. (This room is now the
director's office.) The museo and its director L. M.
Torres did everything possible to make my stay pleasant and to facilitate my work. The fossils were then
under the care of Ángel Cabrera, who soon became a
close friend and helped in all possible ways. He had
been born and grew up in Spain, trained there as a
student of recent mammals, and was already well-
known in that field when he moved to Argentina.
There he found himself in charge of a large collection
of fossils and made many important studies of fossil
mammals as well as continuing some studies of living
South American mammals. He was an excellent artist
and illustrated his own publications. His wife was
hospitable, and they had a nice son who was destined
to become a noted botanist.

Meanwhile, back in New York, Coley Williams had
unpacked our collection and had begun laboratory
preparation of the specimens. On my return there
later in 1931 I had many other things to do, but I
began the study of our collection, combined with the
comparable materials from the two Argentine
museums, now in hand. For the early faunas that I was
beginning to monograph there were at that time three
other, smaller collections to be considered. One of
these was in the Field Museum of Natural History in
Chicago, to which Elmer Riggs had brought back from
Patagonia some specimens from these faunas,
although he had worked more on later strata. The late
Bryan Patterson, subsequently with me at Harvard,
was then in Chicago working over Riggs's collections,
and he arranged for those from the faunas I was
studying to be sent to me on loan. In Paris there was a
small collection in the Muséum d'Histoire Naturelle
made by André Tournouër around the turn of the

century at localities pointed out to him by Carlos Ameghino. Some of the later fossils had been studied by the eminent French paleontologist Albert Gaudry, and in due course I was able to study the earlier ones there. In order to support their work, the Ameghinos had sold a small collection, including specimens from the early faunas, to the university in Munich, Germany. I had visited that institute in my Wanderjahr 1926–1927, and fortunately I had examined and made notes on the specimens of early Patagonian mammals there—"fortunately" because the whole collection was destroyed by bombing during Hitler's war.

In 1933 Scarritt funded a second expedition, and Coley and I again took off to Patagonia by way of Buenos Aires. We did not rework the fossil localities from which we had made collections in 1930–1931, but followed up some leads we had acquired then, and covered areas about which little or nothing was known in the way of fossils. That expedition was on the whole more interesting and fuller of incidents than the first Scarritt expedition, but of course Patagonia was then no longer as impressively fresh and unusual to us. I again recorded it in a long personal journal, but there seemed to be little reason to write another book about Patagonia so soon after the first one. I did, however, include some of the more interesting and unusual episodes in *Concession to the Improbable: An Unconventional Autobiography* (New Haven: Yale University Press, 1978). I need not repeat them here.

When we returned to New York and the American Museum in 1934, Coley Williams continued laboratory work on our Patagonian collections, but in 1935 he found this less congenial and seeing no more interesting possibilities within the museum or in the science of paleontology, he quit and went into a completely different, less demanding but more remunerative occupation. The completion of the preparation of the Patagonian specimens for study, a slow, long task, was turned over to the permanent laboratory staff. As the materials were prepared, I worked over them as I could and wrote some short technical papers about them. I was, however, much distracted by other museum interests and duties. Among these was one that led to the third and last Scarritt expedition in the (northern) summer of 1935.

My own interest, and that of the American Museum since before I was born, continued to concentrate on the earliest faunas (Paleocene and Eocene) of the Age of Mammals, not only in South America, but also and even more intensively and extensively in North America. Much earlier, in 1901 (also before I was born!), a few Paleocene mammals had been discovered in an area east of the Crazy Mountains in central Montana. From 1908 to 1911 a local man, Albert Silberling, who serviced railroad engines during the winter and avidly, successfully sought fossils during the summer, made a stunning collection in that region for the United States Museum (now the Smithsonian Institution's National Museum of Natural History).

James W. Gidley, then that museum's very able paleomammalogist, undertook the preparation and monographing of this collection, but before he could finish that long task he died. It finally fell to my lot to finish what he had started, and then to continue collecting for the American Museum in what had come to be known as the Crazy Mountain Field. Scarritt still had a patron's (or an angel's) interest in fossil collecting and still found me a suitable drinking companion, so he undertook to finance a summer's work there. Al

Silberling and I worked together in that field for four months in 1935, with a camp man and quite a few working visitors from time to time, and one non-working one: Scarritt himself, who had been sub-stantially supporting such work for years but had never seen a fossil quarry. It is sad to report that not long thereafter Horace Scarritt died young, leaving a wife and a young son. Among the memorials to him are the strange extinct beast we found in Patagonia and named *Scarrittia* and a new, important deposit of fossil mammals we found in Montana to which we gave the name Scarritt Quarry, now well known to most paleontologists.

Work on the monograph of the earliest Cenozoic South American fossils continued when possible, but it was much delayed by other necessary projects, in-cluding several books and many shorter studies, a long period of hospitalizations, and by my transferring from the American Museum of Natural History to the Museum of Comparative Zoology at Harvard Uni-versity. A number of other shorter studies on Pata-gonian paleontology and stratigraphy also delayed the completion of the larger project. Nevertheless the monograph planned in 1930 did finally get written, illustrated, and published. Titled *The Beginning of the Age of Mammals in South America*, it appeared as two widely separated volumes of the *Bulletin of the American Museum of Natural History:* volume 91 in 1948 and volume 137 in 1967. An expansion from this start was later written for nontechnical readers: *Splendid Isolation: The Curious History of South American Mammals*, published by Yale University Press in 1980.

There is one other thing to add as a sort of footnote to *Attending Marvels*. On our way to Antarctica in 1970, my wife, Anne Roe, and I stopped over in Comodoro Rivadavia. The now decrepit Hotel Colón still stood, under different management, but there was hardly anything else recognizable remaining from my last sight of the town, now a city, as I left it in 1934. We did not stay in the Colón but in one of the elegant high-rise modern hotels. We drove out to Sarmiento, now not called a colonia, and had a delightful and hilarious reunion with Justino Hernández, now aged and a grandfather, like me. We met his wife, a charming woman of Lithuanian descent as was Justino's mother, and we were joined by Don Casimiro Slapeliz who was only pushing eighty and had so far wrecked only two of his planes. He invited us to take a ride with him in the third, but we politely declined. We were merry after a lunch with ample Mendoza wine, and Justino and I tried to sing some of the songs we sang in camps in the 1930s, such as "Lo' Mendocino' " and "Mañana es Domingo," but for some reason our voices were not up to our memories.

That was the happy part of our travel down the length of Patagonia in 1970. The unhappy part was that we drove hundreds of miles around Trelew, Comodoro, and Río Gallegos on now good roads, and we did not see a single guanaco, mara (called "liebre patagónica" by the local people), or rhea (locally called "avestruz"). In the 1930s we were never out of sight of some or many of these and other Patagonian creatures.

Still more recently, a final touch was put on Patagonia as I had known it when I wrote *Attending Marvels*. On 8 December 1981, the Universidad Nacional de la Patagonia San Juan Bosco was inaugurated in Comodoro.

G. G. Simpson, 1982.